Where Honeybees Thrive

ANIMALIBUS VOL. 10
OF ANIMALS AND CULTURES

Nigel Rothfels, General Editor

Advisory Board:

Steve Baker (University of Central Lancashire)

Susan McHugh (University of New England)

Garry Marvin (Roehampton University)

Kari Weil (Wesleyan University)

Books in the Animalibus series share a fascination with the status and the role of animals in human life. Crossing the humanities and the social sciences to include work in history, anthropology, social and cultural geography, environmental studies, and literary and art criticism, these books ask what thinking about nonhuman animals can teach us about human cultures, about what it means to be human, and about how that meaning might shift across times and places.

OTHER TITLES IN THE SERIES:

Rachel Poliquin, *The Breathless Zoo: Taxidermy and the Cultures of Longing*

Joan B. Landes, Paula Young Lee, and Paul Youngquist, eds., *Gorgeous Beasts: Animal Bodies in Historical Perspective*

Liv Emma Thorsen, Karen A. Rader, and Adam Dodd, eds., *Animals on Display: The Creaturely in Museums, Zoos, and Natural History*

Ann-Janine Morey, *Picturing Dogs, Seeing Ourselves: Vintage American Photographs*

Mary Sanders Pollock, *Storytelling Apes: Primatology Narratives Past and Future*

Ingrid H. Tague, *Animal Companions: Pets and Social Change in Eighteenth-Century Britain*

Dick Blau and Nigel Rothfels, *Elephant House*

Marcus Baynes-Rock, *Among the Bone Eaters: Encounters with Hyenas in Harar*

Monica Mattfeld, *Becoming Centaur: Eighteenth-Century Masculinity and English Horsemanship*

Where Honeybees Thrive

STORIES FROM THE FIELD

Heather Swan

The Pennsylvania State University Press, University Park, Pennsylvania

Library of Congress Cataloging-in-Publication Data

Names: Swan, Heather, 1968– , author.
Title: Where honeybees thrive : stories from the field /
     Heather Swan.
Other titles: Animalibus.
Description: University Park, Pennsylvania : The
     Pennsylvania State University Press, [2017] | Series:
     Animalibus : of animals and cultures | Includes bibli-
     ographical references and index.
Summary: "An interdisciplinary exploration of responses
     to honeybee decline. Features the innovative projects
     of beekeepers, farmers, artists, entomologists, biolo-
     gists, and ecologists in the movement to save the bees
     and work toward a more sustainable co-existence for
     all creatures"—Provided by publisher.
Identifiers: LCCN 2017000150 | ISBN 9780271077413
     (pbk. : alk. paper)
Subjects: LCSH: Colony collapse disorder of honeybees. |
     Honeybee—Effect of pesticides on.
Classification: LCC SF538.5.C65 S93 2017 | DDC
     638/.13—dc23
LC record available at https://lccn.loc.gov/2017000150

Typeset by
Regina Starace

Printed and bound by
Asia Pacific Offset

Composed in
Chaparral & Mrs. Eaves

Printed on
Ensolux Classic

for Elijah and Maia

The western honey bee or European honey bee (*Apis mellifera*) is a species of honey bee. The genus *Apis* is Latin for "bee" and *mellifera* comes from Latin— *melli* "honey" and *ferre* "to bear"—hence the scientific name means "honey-bearing bee."

—GALWAY BEEKEEPERS ASSOCIATION

To bear—11. a. trans. To sustain, support (a weight or strain). . . . 15. To sustain (anything painful or trying); to suffer, endure, pass through. . . . b. To suffer without succumbing, to sustain without giving way, to endure.

—OED

Contents

Acknowledgments

Making this book, like making honey, was an act of collaboration. It takes many honeybees making many trips to many flowers to produce even a teaspoon of honey. Similarly, the writing of this book required the cooperation and assistance of many people to whom I am very grateful. First of all, I'm incredibly grateful to the honeybee people who generously took the time to tell me their stories and share their knowledge. Thank you, Donna and Eugene Woller, Sainath Suryanarayanan, Rufus Isaacs, Dennis and Shelly Hartmann, John and Christina Eisbach, Colleen Bos, Mary Celley, Yin Xianlin, Nathan Schulfer, Ming Hua, Venice Williams, Sharon and Larry Adams, Kaitlin Stack-Whitney, Michael Thiele, and Tang Ya for sharing your expertise. I'm also grateful to the talented and generous artists who agreed to participate in this project: Sibylle Peretti, Elizabeth Goluch, Rose-Lynn Fisher, Aganetha Dyck, Kim Gurney, Lea Bradovich, Sarah Hatton, and Tyler Norman (a member of the Beehive Design Collective).

The book would never have come to fruition without the ceaseless support, wisdom, kindness, and gentle humor of Ronald Wallace, the inspiration and guidance of Anne McClintock and Rob Nixon, the crucial insights of Leslie Bow, and the exemplary work and excellent feedback of Gregg Mitman. I'm also deeply grateful to my brilliant and patient editors, Kendra Boileau and Nigel Rothfels; for the very helpful comments of my exceptional readers, Eric Brown and Jake Kosek; and for the sharp eyes and practiced ears of Alex Vose and Suzanne Wolk. I must also thank Bill Cronon, whose feedback always felt revelatory, Jane Hirshfield, whose gentle advice left my approach to my writing and my life forever altered, Nancy Judd, whose creative commitment to the environment continues to inspire, Keith Woodward, whose sharp wit can always make me laugh, Nancy Langston, whose life and work so closely align with her ideals, and Laurie Sheck, whose encouragement came at exactly the right moment.

I'm endlessly grateful for the friendship and fellowship I found at UW–Madison's Creative Writing Department and the Nelson Institute's Center for Culture, History, and the Environment; the latter awarded me a fellowship that helped me make my trip to China. Thank you also to James Crews, Sarah Nelson, Peter Boger, Jenny Seifert, Jennifer Boyden, Mary Fiorenza, Amaud Jamaul Johnson, and Nathan Jandl for believing in my work and for helping me hone my voice. Thank you, Marc Basch, for being my writing pal from the very beginning. Thank you, Deb Sullivan, for reminding me to keep swimming. Thank you, Jeanie Garvey, for your beautiful energy. Thank you, Erika Ono, for always knocking on the door. Thank you, Robyn Shanahan, for your consistent support. Thank you, Andrew Mahlstedt, for many things. And thank you, Matt Rosenthal, for continuing to navigate all that we do. Thank you also to my students, D. J. Gillins and Dcota Dixon, who taught me as much as or more than I taught them. And thank you, Anna, Rick, Jackson, Donovan, and Benjamin Waters for the love and the "just do it" speeches.

This book would never have been written without my amazing neighbors, who taught me the real meaning of *It takes a village*: so a huge thanks to the bird whisperer Heather Beasley and her fish expert husband, Pete McIntyre; to the guru of coffee and mindfulness J. J. Kilmer; to the generous and creative Tyska family—Leane, Steve, Esme, Franny, and Sylvie; to the truly wonderful Fran Nelson; and to my personal superheroes, Susan and Tom Willis, and their children, Josephine, Jack, and Ike. And, of course, I must thank the honeybees, for obvious reasons. I would not be a person who even noticed bees or light or poetry or art if not for my beauty-filled parents, Stephanie O'Shaughnessy and Keith Davis, who raised me in art studios and gardens and who continue to inspire and guide me to this day. And, finally, thank you, Elijah and Maia, my remarkable children, for your love and your light and your humor, and for reminding me daily that life on this precious planet is a gift worth fighting for.

Introduction

BEES AS INDICATOR SPECIES

Thursday, 6:30 A.M. Coffee in one hand, bee veil and hive tool in the other, and on the horizon, the peach-stained promise of sun. Mornings are opening later now. The brisk air reminds me of what is to come . . . the merciless cold of a Wisconsin winter. But today the forecast predicts a warm afternoon. For a long while, I trudge upward through the trees and then across a meadow through undulating mist. The goldenrod is just past full bloom now, and the remaining pollen dusts my pant legs. I don my veil at the fence and am plunged back into semidarkness. Once inside the gate, I can hear the box humming. I freeze at the sound of something scurrying nearby, sticks breaking. Coyote? I wonder. But what falters in me momentarily is soothed by the murmuring of the hive, the lulling sound of honeybees.

I haven't used the smoker much this year, though as an apprentice I learned to pack it with leaves and wood chips in just the right combination to keep a fire smoldering inside the coffee-can-sized container. I've found that opening the hives without it, ever so slowly and carefully, seems to upset the bees less. This morning is no different. I gently crack the hive cover and then carefully pry off the inner cover. The smell of propolis and beeswax fills the air. Inside is a seething metropolis. Approximately sixty thousand bees live here. A familiar frisson fills me. The frames are crawling with tiny creatures whose tongues are encyclopedias of plants, whose bodies are an unfolding cartography, whose wings bear the weight of a global food system—and whose disappearance is a message to the world.

The way bees convey messages to one another seems almost magical. One worker bee crawls down through her wax chambers to the hive opening and takes flight on her tiny wings, then moves through the richly scented air until she is lured by some particularly delicious smell—say, a blossoming pear. She gravitates to a single flower, lands lightly on the petals, and then probes the center of the blossom. She drinks in the sweet nectar and packs nutritious, protein-filled pollen onto her legs. After visiting many blossoms, she uses the sun as her guide to return to the hive, where she shares her riches. But sharing pollen isn't all she does. She also shares the location of her discovery. Her body begins to quiver, and she dances a map to bees nearby. The other bees decode the cartographic dance and become familiar with the scent of the pear blossom, which lingers on her body. With this information, her hive mates can precisely locate and identify the same blossoms in the same orchard and make their own visit to the pear trees. It is this ingenious method of communication that has allowed humans to use bees as pollination assistants for centuries. But recently, their incredible olfactory powers have been put to other uses.

A honeybee can be trained to detect drugs, disease, bombs, and minerals, all by smell alone. A bee reliably extends her proboscis when she associates a specific scent with a reward. If, for instance, you feed her some sugar syrup laced with a chemical, she will extend her tongue when she encounters the smell of that chemical again. Her sense of smell is said to be thousands of times more sensitive than that of a human or even a dog. Sniffer bee experts place bees in tiny harnesses, securing their heads and arms in a contraption (not unlike the punitive stocks of the Middle Ages) inside an airtight chamber. A tiny camera focuses on the bees' faces. With the bees in position, the researchers can introduce a quantity of air into the chamber and determine whether a specific scent is present by watching to see whether the bees extend their tongues. The accuracy of this bee detection system has led researchers to explore a variety of uses for the bees. The Berkeley geographer and beekeeper Jake Kosek has written extensively about the use of the honeybee in the military, for example. Many find that they are useful as bio-indicators.

At Inscentinel, a company located just north of London, a group of scientists and beekeepers have developed a small machine that looks a bit like a Dustbuster, with bees inside, that can be used in airports for sniffing out drugs or explosives. A designer named Susana Soares has created some glass bauble devices that use bees for detecting disease. Bees have successfully detected tuberculosis, cancer, and diabetes. In Asturias, Spain, a mining company hoping to reopen a gold mine used bees to detect levels of heavy metals in the environment. The bees' pollen was tested after they fed on plants that get their nutrients from the soil and water surrounding the mine; the pollen tests were one of several environmental samples the company used to prove to local residents that the mine was safe. But what else are bees telling us? Since 2004, bees and other pollinators have been disappearing at astronomical rates. Are honeybees acting as the canary in the coal mine?

I insert my hive tool into the sticky, propolis-covered cracks between the frames. With my fingers pulling gently and steadily, the frame comes slowly free. The sun is streaming through the trees now, so I lift the frame up to the light. Behind the bodies of the bees, I see that the lattice of wax vessels is filled with luminous honey. From a tear in the wax oozes a rivulet of the gold liquid, the sweetest substance in the world. Farther down in the hive body, I see that the queen is slowing her brood production. The darkened cells, cradling baby bees, are surrounded by chambers packed with pollen. "It takes a cell full of honey and a cell full of pollen to make a bee," says Eugene Woller, one of my favorite beekeepers. I hope the pollen and the nectar they've gathered are clean. Pesticides, even in very small doses, are devastating to bee health, especially affecting their keen senses. And worldwide, many agree, bees are dying, in part, because of the chemicals we use on our crops and our lawns.

As early as 2004, bees began disappearing in staggering numbers. Millions of bees died suddenly, and thousands of hives failed. Beekeepers and scientists around the world scrambled to find an answer to what became known as Colony Collapse Disorder. Bees would disappear overnight, leaving no trace, the baby bees left to fend for themselves. For several years, competing theories emerged. Was it cell phone radiation? New viruses? Mites? Poor nutrition from habitat destruction? Stress from being shipped around from crop to crop on the back of semi-trailer trucks? Or pesticides? Perhaps all of these things were contributing. By 2011, the United Nations had made a statement echoing the mythologized words of Albert Einstein, who allegedly said that without bees, the human race would not last longer than four years. The UN made this claim on March 10, 2011: "The *potentially disastrous* decline in bees, a vital pollinating element in food production for the growing global population, is likely to continue unless humans profoundly change their ways, from the use of insecticides to air pollution" (emphasis added). Potentially disastrous, the report explained, because bees and other native pollinators are integral to the production of at least one-third of all food eaten globally. Almonds, coffee, berries, pumpkins, herbs—even dairy and meat, as cows eat clover—all require pollination by insects.

Human reliance on insect pollination has made the bee the darling of agriculture for centuries. Bees have been moved from crop to crop since 2500 B.C.E. in Egypt, when, the hieroglyphs suggest, Egyptians moved bees up and down the Nile. Today, truckloads of bees travel to huge monocrops in order to pollinate more efficiently. The almond groves in California use 1.6 million beehives for their pollination season. The acreage for that crop is about the size of Rhode Island and produces approximately two billion pounds of almonds annually. The bees are shipped between orange groves in Florida and cranberry bogs in Wisconsin as well. These crops are covered with herbicides and pesticides to increase crop yields. Commercial pollinators feed the bees high fructose corn syrup laden with antibiotics, a concoction intended to keep them energetic and healthy. But this is not the result. A USDA study released in 2014 has proved that bees fed artificial diets are much more vulnerable to damage from pesticides. Far from thriving under these conditions, bees are suffering, but their suffering has not been on a scale immediately detectable to our distracted culture.

The bees and many other pollinators have been victims of what Rob Nixon calls "slow violence." "We must address our inattention to environmental calamities that have staying power," Nixon writes, "calamities that patiently dispense their devastation while remaining outside our flickering attention spans—and outside the purview of a spectacle-powered media." In 2006, however, the bee crisis did reach spectacular proportions. The hectic pace imposed on the bees and the decades' worth of incremental violence against the species finally began to take their toll. Bees began dying by the millions. As scientists and beekeepers searched for answers, the bee began to gain some public attention. Newspapers suddenly began reporting on Colony Collapse Disorder. Without bees, there would be very little food. This was an important problem.

For several years, scientists, farmers, and beekeepers were busy asking questions, observing die-offs, and trying to find answers. Many beekeepers argued that pesticides were the culprit, but proving this was more difficult than might be expected.

We live in an era in which many people privilege scientific knowledge over all other kinds of knowledge. "Scientists have proved. . . ." "New research shows. . . ." These phrases are what a good portion of our society trusts. Certainly, scientific research is absolutely critical to understanding the multiple environmental challenges this planet is currently facing, climate change being the most obvious and pressing of these. At times, however, reliance on science can be problematic, and the trouble seems to be twofold. The first issue is that knowledge formed outside the scientific realm is often considered dubious or unfounded. The knowledge of plants among people indigenous to an area of rainforest, for example, might not be taken seriously or even heard at all, and frankly, this kind of knowledge is often lost as corporate enterprises usurp power and take land from local peoples. The knowledge of a beekeeper who has spent years observing bees and bee behavior might not be given due respect, because this knowledge hasn't been proved by science. Sainath Suryanarayanan and Daniel Kleinman have done much work on this particular issue. Suryanarayanan, an entomologist and science studies scholar who has been working with wasps and honeybees for years, suggests that the ways in which we ask questions or search for scientific proof of something may be too narrow. For example, you can devise an experiment to test an increasing quantity of a specific pesticide on a group of bees and watch to see how much of the chemical it will take to kill them. But what if the chemical does not affect the bees immediately? What if its effects are more subtle and secretive? And how might you test for the innumerable combinations of pesticides that can end up in a bee's body? In an afternoon, she might ingest a cocktail of ten different substances. How do you test for that scenario? Suryanarayanan, whose work I explore in chapter 4 of this book, suggests that we must change the way in which we ask the questions and which voices we choose to listen to.

The second concern about our reliance on science is that sound studies can be called into question simply by casting doubt. Creating doubt about the certainty of a scientific conclusion can dramatically affect public opinion and policy. The power of this strategy is illustrated in the book *Merchants of Doubt*, by Naomi Oreskes and Erik M. Conway. "The U.S. scientific community," says the website advertising the book, "has long led the world in research on public health, environmental science, and other issues affecting the quality of life. Our scientists have produced landmark studies on the dangers of DDT, tobacco smoke, acid rain, and global warming. But at the same time, a small yet potent subset of this community leads the world in vehement denial of these dangers." How could scientists and beekeepers prove that bees were the victims of pesticides when huge corporations like Bayer and Monsanto produced and supported the use of these chemicals? Somehow, though, the truth usually comes out in the end. The truth about DDT was exposed, for example, but problematic new man-made chemicals followed.

The bees we are worried about have also been created by human manipulation. Like cows, dogs, chickens, pigs, and rabbits, bees have been genetically engineered for our purposes for many years. As Jake Kosek has shown in his work on the critical natural history of the industrialization of the honeybee, the bee has been genetically manipulated to be hairier, to carry more pollen, to be more disease resistant, to be more docile. In the past, queen bees mated with many drones, but today most queens are artificially inseminated. Rudolf Steiner, among others, argues that these manipulations have created a weaker bee. And while we have been changing the bee, the chemical industry has been brewing more efficient pesticides and herbicides in the name of higher crop yields. The plants themselves have been changed as well. So we have created this crisis. But, sadly, it does not affect only us. It does not affect only our honey yields and the pollination of our crops. These practices have also created a problem for all the other pollinators, have poisoned our water tables, have killed our birds. In 1962, Rachel Carson's *Silent Spring* sounded a clarion call. DDT was destroying the land, the water, the birds, and the insects, and would eventually destroy humans, too. What pesticides would she write about today?

While I was writing this book, newspaper and magazine headlines announced that scientists had determined the cause of colony collapse. Reuters, for example, announced, "Mystery of the Disappearing Bees: Solved!" Neonicotinoids, a neurotoxin not unlike DDT, were identified as the culprit: "In the U.S. alone, these pesticides, produced primarily by the German chemical giant Bayer and known as 'neonics' for short, coat a massive 142 million acres of corn, wheat, soy and cotton seeds. They are also a common ingredient in home gardening products." But while one part of the problem has been named, changing our behavior happens slowly, and this is not the only problem that bees and other pollinators face; rather, they confront a constellation of issues. One or two news articles will not change agricultural and cultural practices. And, unfortunately, there is plenty of news that clouds these findings.

In 2012, Facebook was abuzz with news of a Stanford study on the benefits of organic foods. The *New York Times* reported that organic food was really no different in its nutritional content from food that was conventionally grown. Critics quickly flooded the Internet with rebuttals. But how many readers never questioned this finding, and instead felt validated in their choice to continue buying pesticide-laden foods? Unfortunately, the article did not address the real issue, which was not whether organic heirloom tomatoes contain more vitamin C than hybrids but that, once again, the public was being told that it was okay to eat food treated with toxic pesticides, which enter not only our bodies and the bodies of farm laborers but also our soil, water, and air—and, of course, the smaller bodies of animals and insects. The great irony is that if we kill off the bees for the short-term gain of faster-growing plants or larger yields, we will soon enough be without food altogether because we'll have no pollinators. We need the bees in order to survive.

I begin to feel a habitual pang of guilt for invading the hive, so I look at a few more frames and then replace the hive cover. There is no sign of any disease or fungus or mite problem today. Beekeepers have become accustomed to looking for those things now. After a ten-thousand-year relationship with humans, the bees are not thriving, in large part because we have cut down the fields of wildflowers and poured toxins onto our crops (human folly!), and therefore it is humans who must now choose to save the bees.

This is not, as must be evident by now, an entirely unbiased book. It is part love song, part lament, part quest. Because I am keenly aware of many of the problems bees are facing, I have been on a quest to find people who are actively and creatively confronting and addressing the issue of pollinator decline, while trying always to be aware of the complexity of these issues and the lack of a simple solution. In these pages, you will be opened to the passion and the unique work of beekeepers, entomologists, artists, poets, farmers, and ecologists from all over the world who have at their heart the desire to heal the bees, not only because they make fruit and flowers and honey possible, but because they are unique, incredible creatures. Because this book focuses on individual people working for change, interviews have played a large part in it. I traveled to China, where I spent time with Chinese beekeepers in an area near Chengdu where most bees have been wiped out by pesticide use, and the fruit growers have been forced to pollinate by hand. I interviewed blueberry farmers in Michigan who are exploring integrated pest management with the help of entomologist Rufus Isaacs, in order to provide more food for pollinators and also to encourage "beneficial" insects who prey upon crop pests. I also worked with urban farmers in Milwaukee, Wisconsin, who use the bee as an image of their resistance communities and raise their bees without chemicals. The brilliant artists I feature in the "gallery" sections of this book offer unique perspectives on honeybees; they include a South African artist, Kim Gurney, whose amazing work in response to the bee crisis pays homage to the indigenous San honey hunters, and a Canadian artist, Aganetha Dyck, who enlists the cooperation of live bees to make her artwork. Another central figure in my project is the entomologist Sainath Suryanarayanan, whose work with bees involved pouring pesticides into hives and slowly killing them in order to see how much of each chemical the bees could endure. The work so haunted him that he shifted his career. He is now exploring innovative, interdisciplinary approaches to science that would be more effective and more "humane" to bees. While my book clearly dramatizes the dire situation for bees, it also illuminates the many innovative projects under way worldwide that are attempting to transform our unsustainable practices. The fate of honeybees is precarious, but, as this book makes clear, the embattled bees now have an extraordinary collection of imaginative allies, from scientists and apiarists to artists.

A couple of years ago, in northern California, I took a walk with Michael Thiele, a beekeeper who believes very strongly that we need to listen to what the bees are telling us. The bees are dying because we have been using them as tools in an industrial landscape, he said. We stood by one of the hives he'd made in a hollow log. He told me that they understand something we humans may not, that their "life body" extends beyond their individual body and depends upon the larger community. This beautiful, harmonious, inter-being way of living is one that humans need to emulate, he said. The bees take care of one another and their environment. And we've forced them into "unnatural" ways of life. Michael and many other beekeepers around the world are trying to create new ways of being with bees to ensure their survival—and the survival of humans as well. In a shed, he showed me another kind of hive he'd designed that would allow bees to build large, semicircular layers of comb, as they would if they were building in a tree. The Langstroth hive design was not designed with a bee in mind but for human convenience. For Thiele, beekeeping is not about honey production or massive pollination projects but about the health of the bees. For him, the bees are spiritual teachers. Indeed, in their presence, I feel humbled and quiet . . . and then very sad. Why is it that our human enterprise must be so destructive? How can we move toward a more sustainable coexistence? How can we restructure our lives so that we are less destructive, to ourselves and to them?

The sun has burned off the mist, and the field is dazzling now. The birds are loud. I'm not sure whether it was the flavor and smell of honey that first inspired my passion for bees or if it was the discovery of their beautiful careers, flying among flowers, taking care of the plants and the baby bees; but I've been enchanted with them for as long as I can remember. I have always loved watching their intricate bodies move among petals, watching them work fields like this one. I remember discovering Virgil, who was also intoxicated by bees, who also loved to watch them and had suggestions for anyone who wanted to build a home for them. He knew which flowers they liked, and said they preferred a home with the sound of a bubbling brook nearby. Like me, and so many others, he found them worth thinking and writing about. It seems like a privilege to share the world with creatures like these.

I carry my veil and my empty cup back through the woods. I like to think there is a chance for these insects. Today, at least, in one hidden city, the honeybees are still thriving.

Honey Business

chapter one

> If I die in a bee yard with a hive tool in
> my hand, I'll die a happy man.
>
> —EUGENE WOLLER, BEEKEEPER

Many spring mornings, I have biked to the Madison, Wisconsin, farmers' market early, 6:30, before the crowds. Folks are usually still setting up. There's a banjo player just beginning to pluck out a quiet tune, bakers from Stella's unloading big trays of cinnamon rolls, sellers arranging sunflowers and zinnias, farmers lining up boxes of tomatoes and kale. When I want to pick up a big chunk of beeswax, I find Eugene Woller, a beekeeper friend of mine. Eugene and his wife, Donna, have been in the honey business for forty years. Every Saturday, from April to November, they set up their stand at the farmers' market. In the early days, Donna told me over coffee one afternoon, they would bring their kids and the honey up to the square in a little wagon. Nowadays, they have delivery trucks and a simple but handsome honey stand with their logo on the side. At one end of their front counter, a tall glass case houses a miniature center of industry—an observation hive filled with dark comb and live bees. Both children and adults stop to stare, mesmerized by the strange architecture and the activity of magic taking place in the secret chambers of the hive. These tiny beings—who have been said through the ages to carry souls to heaven, to make food worthy of consumption by the gods of many traditions, whose quiet, industrious communities have been held up by writers across time as models for human society, these ingenious insects—are right there on the counter for anyone to observe. How can these insects really create the sweet gold displayed at Eugene's stand? Eugene shares the story of bees with all who are willing to listen. And he and his family have been good enough to share their story of a life with bees with me as well. One of my most memorable visits with them occurred during the honey harvest.

———————

The roads to Eugene's honey house wind through the hills of southern Wisconsin like grape vines. The day I was visiting, the hills and valleys offered a feast of color. The oak leaves had turned burnt orange, and the road was edged with the deep reds of sumac, the burnished yellow of goldenrod, the occasional purple of asters; above, in the cobalt blue sky, red-tailed hawks made great arcs as they surveyed everything below. I passed graying barns and old silos and an occasional small field of corn or alfalfa. Most of these small farms were once family-owned dairies that Eugene says were put out of business by big corporate dairies. But all I was thinking about was how this must be a great place for bees.

Eugene's enterprise is tucked into the south-facing side of a hill that looks out over one of the valleys outside Mount Horeb, Wisconsin. Two warehouses and a tin roof over a wood-burning furnace sit closest to the road. A curl of smoke rose from the wood burner, which heats his home and the honey house all winter long. Eugene and Donna and their black lab, Buddy, live in a little house hidden higher on the hillside among a stand of oak and aspen. Behind the warehouses are a little group of white Langstroth-style hives, all leaning slightly to the left or right, like a group of friends laughing at a good joke.

When I stepped inside the honey house, I was enfolded in the smell of honey and wax. I nearly swooned. When I was a little girl, my father took me to the home of his friend Bob Baker, where we participated in a honey harvest outside Mineral Point, Wisconsin, and since that time I've been in love with bees and their sweet product. In Hilda M. Ransome's marvelous book *The Sacred Bee in Ancient Times and Folklore*, I had read that honey was extremely useful when trying to strike a bargain with the gods. To honor the goddesses called the Eumenides, for example, once a year, in a grove of evergreens, a festival took place that involved ceremoniously pouring libations of honey mixed with water and flowers. Honey cakes were used to appease the menacing three-headed dog Cerberus, who guarded the gates of Hades. In the homes of ancient Greece, on the third and fifth days after a baby was born, a small bowl of honey, three almonds, a loaf of bread, and a cup of water were left out on a table to please the Fates. Honey was also fed to the sacred snake that lived on the Acropolis; and among the people of Epiros, to ensure a good relationship with Apollo, snakes were fed honey cakes by the hands of naked maiden priestesses.

But no one else in Eugene's honey house was swaying with romance or spiritual reverence. Harold, a neighbor and an old friend of Eugene's, was hunkered over a few six-by-six-inch frames of comb honey. With a little knife, he was mindfully scraping the wax off the edges of the wood so the comb would be ready for market. Jack was moving empty Ball jars into position to be filled at the spigot. Every jar is filled by hand here. Carol was busy arranging Eugene's delivery schedule on a whiteboard. One week, Chicago; the next week, Green Bay; and the coming Friday, a trip to talk to the Mount Horeb second graders. Danny, the newest and youngest of the help, was out in the garage loading the flatbed pickup with "hive bodies"—what they call the boxes the bees live in. I found Eugene in his olive-green coveralls beside a stack of honey barrels getting some spun honey started. "I've only been keeping bees since 1965," Eugene had told me when we first met. He was the "rookie" compared to his mentor and good friend, Emmett, who had been keeping them since the age of five, and Emmett was in his nineties. The two of them met at the University of Wisconsin–Madison in a bee course hosted by the Department of Entomology. Eugene has kept bees ever since.

He greeted me with his usual energetic warmth and suggested that I get some coveralls on myself. Once I was suited up, the two of us helped Danny finish loading the truck. "We'll be pulling honey and replacing wets today," he told me. "Wets," I would learn, are the honey supers—the top boxes of the hive—filled with frames of emptied honeycomb. The honey was extracted here at Eugene's honey house in large extractors. The bees cap each wax cell of honey, and the frame of honeycomb is essentially like many tiny wax jars of honey that humans, for centuries, have liked to open and eat. Uncapping is generally done with hot blades that are run along the wax tops of the cylinders. The wax from the caps is saved and melted down for candles and soap. The open comb is placed, with many other oozing frames, like spokes on a wheel, inside the large metal drum of an extractor. When you turn on the machine, it whirls the frames around and around, and the centrifugal force draws the honey outward to the steel walls of the drum. Gravity does its work, and the gold liquid pours down the sides of the drum and into a spout. From there, depending on how much honey you are harvesting, you can fill jars or barrels of sweetness. Once the comb is empty, though, it's best to return those wax combs to the bees, who, Eugene explained, will clean them up and reuse them.

The three of us squeezed into the cab of the truck, our veils and hive tools and gloves resting on the dashboard. Eugene was hoping that yesterday's rain wouldn't make the drive out to the bee yards too muddy. The lines of Eugene's face are etchings of a joyful life. He squinted out at the gorgeous day we had been given, and we were off.

The first stop was an orchard full of apples and pears. Eugene said that the farmer was interested in a style of tree pruning that originated in New Zealand. The trees were short, sturdy, and fully leafed out, and while I was not exactly sure what made them unique in their shape, I did notice that all the branches seemed to be reaching for the ground. The apple growers had had a tough year. A heat wave brought an abrupt end to winter in Wisconsin, and the trees blossomed nearly a month early. An orchard owner west of Eugene's place, near Spring Green, Wisconsin, told me he remembered that time very clearly. On April 17 and 21 the temperature dropped, and there was a frost. Most of the open blossoms were killed, but most of those that were still closed like little fists made it. Whether they were open or not depended on the species, and whether they were up on a hill in the sun or down in the valley in the shade affected how cold they got. That farmer had lost 50 percent of his crop. My neighbor and I bought two bushels of the survivors and made some amazing cider. Eugene said that this farmer had also not seen a large crop this year. How were the bees, then? The bees, apparently, had done just fine.

The effects of climate change will require some thoughtful responses from farmers in this region. Whether climate change brought the early spring is a matter of contentious debate, but the general warming trend is something that all the farmers are aware of, and managing it will present many challenges. That summer, the Midwest had endured one of the worst droughts in a long while. Strangely, however, the drought had actually benefited the bees. Bees love alfalfa blossoms, and because the last crop of alfalfa never grew tall enough to warrant harvesting, the bees had an extra couple of weeks of blossoms. Eugene cringed with what appeared to be some guilt as he said, "The truth is that it was great for the bees."

Our first stop was a quick one. There were probably fourteen hives, bees buzzing slowly in and out. It was not warm enough for them to move very quickly at that point in the morning, at least until we bothered them. We removed the metal hive covers and stacked a "wet" on each of several of the hives before hopping back in the pickup. "We'll be taking honey out of this next bee yard."

Once again we were winding through hills, farmland, and occasionally a little unincorporated town, like Springdale, which now consists of a few farms and an old town hall. I was so glad Gene was driving. There were so many little turns on unmarked county roads, or, if they were marked, I still could not keep them straight: County G, County JG, County A, County Z. Suddenly, he was turning off the main road onto a dirt road, which was really just two muddy ruts a pickup truck's width apart, leading through the weeds and into a stand of birch and maple trees. This was why Eugene was worried about the rain, but he did not hesitate. The truck slowly lurched forward, and we were suddenly driving through the forest. "Funny there aren't any deer today," Eugene said. "I usually see a family of them in here." He pulled up over a hill and stopped the engine. "Here we are!" he chirped. We had stopped in a tiny meadow overlooking a valley full of autumn color. Two lines of beehives in various shades of white stood in the sunlight. It was perhaps a place like this that Yeats had imagined when he wrote:

I will arise and go now, and go to Innisfree,
And a small cabin build there, of clay and wattles made;
Nine bean-rows will I have there, a hive for the honey-bee,
And live alone in the bee-loud glade.

"Time to suit up!" Gene shouted, startling me out of my reverie. Oh, yes, this was work; I had to remember. I pulled the long strings to tighten my bee veil—a transparent curtain that hung from the rim of my hat—and tied them around my waist. Eugene helped me pull on the long leather gloves. Novice that I was, I wondered at the necessity of so much protection because I hadn't been stung much before, but I would quickly realize that this was a hobby beekeeper's delusion. When you are pulling honey from many, many hives with the method we would use, this uniform was essential. That is, for me it was essential. Eugene went through the whole process bare-handed.

My job was to fill the smoker. Eugene reached down behind the rear wheel well of the pickup and pulled an old steel box out from a shelf made of a couple of pieces of iron. The smoker was stored inside this box to prevent fire as Eugene moved between bee yards. I set about building a tiny fire in the metal cylinder of the smoker with bits of cardboard, as Eugene and Danny pulled a small generator off the bed of the truck. "I don't like chemicals," Eugene said. "Some guys use 'em, but I like to do it this way." I was not sure how one would use chemicals to remove bees, and when I later looked into it, I learned that you can put a variety of bee repellents, such as benzaldehyde, into the hive to make the bees evacuate. Knowing even the little I did about a bee's astonishing sense of smell, I could imagine how offensive that might seem to a bee. And Eugene loved bees. He hooked up to the generator a blower with a long tube like the attachment for a vacuum cleaner. I realized that we were going to blow the bees off the comb. That meant a lot of mad bees in the air. I was suddenly glad I was wearing gloves. "We'll need a lot of smoke," Gene said.

The process began. We had almost twenty hives to check, and so we moved fast. At each hive we removed the outer hive cover, popped open the inner hive cover, which was often glued in place with a sticky substance called propolis, and pulled it off. Inside, the bees were busy, hundreds crawling across the top of the sticky frames. "Smoke 'em," Eugene said, and I pumped the bellows on the side of the smoker, and a puff of smoke billowed out. The bees instinctively responded to the smoke by going down into the hive to fill their bellies with honey. The hive boxes on top of the stacks were the ones filled with honey frames. The lower boxes contained brood chambers filled with baby bees. These were left intact. It was the top boxes, or "honey supers," that we would put on the truck once the bees were blown off. Eugene decided as we went along which hives had just enough honey for the winter and which had a surplus for him to harvest. I was amazed at how often he took frames full of honey from his own stores and put them back into the hives for the bees. This proved to me that he had their health and best interests in mind. He was willing to lose a little honey money to keep his bees alive. Some beekeepers would supplement the bees' diet with high fructose corn syrup, but Eugene was leaving enough honey for the bees to eat throughout the winter.

To remove the bees, Eugene flipped the super on its side so that the frames were all exposed. The frames are separated by what is called a "bee space"—room enough for bees to do their work. Once the super was on its side, he started the blower and aimed the stream of air between the frames.

I'm actually not really afraid of bees, but suddenly we were standing in a cloud of thousands of them who had all just been blown out of their home by a gale-force wind. A wave of fear and awe spread through me. I looked down at Eugene's huge, muscular hands and realized that he was being stung over and over. He didn't flinch. He kept going, completely undisturbed. Once again, I was surprised. A bee sting is not insignificant. My own sting experience is one of an intense burning sensation that radiates outward, the way a stone thrown into water will create concentric waves. Once, after a poor bee had crawled up my foolishly loose pant leg and I had crushed her between my thigh and my calf when I crouched down—causing her, appropriately, to sting me—I had a hot, swollen circle the size of a hydrangea blossom for three days afterward. And I'm not even allergic. According to the National Institutes of Health, 3 percent of the population can suffer from anaphylaxis or an allergic reaction after an insect sting. When you experience anaphylaxis, as Eugene's wife, Donna, did, early in their relationship with bees, your throat and tongue swell and make breathing difficult, your heart rate increases, and you can end up unconscious if you don't receive treatment in a hurry. Realizing that her life-threatening reaction to stings was going to make life very difficult, Donna sought out a doctor who treated her with bee venom once a month until she developed immunity. Eugene certainly seemed immune, even to the pain.

For those who are not allergic, however, bee venom is considered beneficial. The unique composition of enzymes, peptides, phospholipids, and other compounds is thought to cure and prevent all kinds of maladies. Beliefs about bee-venom therapy span the globe and appear in texts on Greek and Chinese medicine. Pliny the Elder claimed that bee venom could even prevent baldness (though Eugene's head suggested otherwise). Today, sufferers of arthritis or multiple sclerosis can seek out therapeutic stings to alleviate the pain associated with those diseases. Very recently, some evidence has suggested that the venom of the honeybee might even be able to destroy cancer cells.

Standing amid the cloud of bees, I soon realized that the bees flying around us seemed to be more startled than angry. I relaxed a little and for a moment stood still, hypnotized by the amazing number of tiny whirring bodies. But we had a lot of work to do, so I got back to business. Danny carried boxes of honey to the truck and Eugene moved through the hives, while I worked the smoker and closed the hives back up after him. Within an hour, we had completed the yard. Before we left, Eugene dipped a hive tool into a honey-filled comb and handed it to me. "Try a little taste." I put a honey-covered finger into my mouth. Nothing sweeter in the world.

There are many different kinds of honey, though, as Eugene was quick to remind me. "You got orange blossom honey, tupelo honey, almond honey, honey that comes off of cucumbers, you know, and they all have a slightly different taste." Buckwheat honey is like a stout, the color of walnut wood and with a flavor as rich as molasses. The honey we were eating had come mostly from alfalfa and some late-season wildflowers, like goldenrod. And it's true: honey tastes like flowers on the tongue. Its flavor directly reflects the kind of nectar that the bee has been sipping.

In the spring of 2012, at Allen Walker's farm in southwest Florida, I sampled black mangrove honey and tupelo honey. Walker told me that there were more than three hundred kinds of honey in the United States alone. His bees enjoyed orange, black mangrove, and saw palmetto blossoms primarily, depending upon the season. He maintained that the mineral content of the soil also contributed to flavor, as some soils are sandy, some are clay-based, and some are primarily composed of humus. Black mangroves grow in saltwater, and I swear I could taste the saltiness.

The color of honey also changes in response to the daily diet of a bee. Buckwheat honey is a deep chocolatey brown, while white clover honey is like bottled golden sunlight. In 2010, in Brooklyn, New York, beekeepers were perplexed by the discovery of shockingly bright red honey in their hives. No one had ever seen anything like it. Careful detective work led them to an unusual source of sweetness: a maraschino cherry factory. The *New York Times* mused, "It seems natural, by now, for humans to prefer the unnatural, as if we ourselves had been genetically modified to choose artificially flavored strawberry candy over strawberries, or crunchy orange 'cheese' puffs over a piece of actual cheese. But when bees make the same choice, it feels like a betrayal to our sense of how nature should work. Shouldn't they know better?" But the ample, cloyingly sweet maraschino juice was hard for the bees to resist, and the result was honey dyed bright red by FD&C Red Dye #40. Similarly, candy-craving bees in Ribeauvillé, France, began producing dark green and blue honey when they discovered they had access to an M&M factory. Sometimes traditional blossoms create surprising results, too. Beekeepers on the Madison, Wisconsin, beekeepers' listserv have described honey from purple loosestrife that tasted, according to one beekeeper, like Juicy Fruit chewing gum. Another person wrote about a curious batch of honey that tasted like Jolly Rancher watermelon candy. Our boundaries between natural and unnatural are always being challenged these days.

With a pickup truck full of southern Wisconsin honey, we piled back into the truck and went on our way. Eugene took us next to his favorite lunch spot—a tiny, one-room white clapboard church that sits on one of the highest hills around. The three of us sat on a picnic table in the sunlight, a light breeze playing in the golden leaves above us. Our bodies were warm with work, and sitting felt good. The landscape of southern Wisconsin spread out in all directions. The industry of growing things was evident in the quiltlike patterns below us. The story of Eugene's beekeeping cannot be told without telling the story of Wisconsin agriculture.

"Things have changed," Eugene said to me in one of our earliest meetings. "Our crops have changed. Our farming methods have changed. . . . Things have changed a lot." Over the years, Eugene has seen small farms sell out to bigger ones. Small-scale dairy farmers who grew fields of alfalfa for their cows eventually sold those cows and switched to cash crops like soybeans and corn. Along with the cash crops came GMOs, new pesticides, and chemical fertilizers. "The problem that comes in with spraying, especially aerial spray, is that the stuff gets on the leaf. The bees pick it up, drag it back to the hive." Bee bodies are designed to collect everything they can from flowers. They not only drink nectar; they attract pollen. Their entire bodies are covered with fine hair that enables them to collect dust and distribute it from flower to flower. But this is not an unselfish or an accidental act. Bees need the protein from pollen and the energy from nectar to survive. Pollen and nectar, as well as leaves and roots, are filled with human-made chemicals these days. The pesticides are bred into the plant, so they exist in every part.

Eugene's dislike of chemical pesticides seems quite natural given the era in which he grew up and his long-term relationship with bees. Eugene and his mentor, Emmett, were at the University of Wisconsin–Madison in the mid-1960s, when there was a great shift in environmental and agricultural thinking. Aldo Leopold had published *Sand County Almanac* in 1949, inviting his readers to think about his "land ethic," which posited them in the center of a community that included "soils, waters, plants, and animals" and paved the way for ecologically minded writers. In 1962, Rachel Carson published *Silent Spring*, her creative non-fiction piece about the dangers of DDT. Her aim was to write a book based on strong scientific evidence that the public could understand, and she succeeded wildly; the book became a groundbreaking classic. After World War II, many new pesticides were developed that were thought to be miraculous solutions to the problems of agricultural pest management and insect-borne diseases like malaria. What were not clear were the effects on the ecosystem and the bodies of humans and nonhumans alike.

But even before the publication of *Silent Spring*, consciousness about the deadly effects of DDT was increasing, and the public began to mobilize. Citizens saw the effects of the chemical in their daily lives. In the days before the Internet, organizing large groups of individuals around an issue took a great deal of passion and persistence. In 1957, Mrs. F. L. "Dixie" Larkin, the chairwoman of the Committee of a Thousand—representing a thousand organizations concerned about the risks of DDT—wrote a letter to the Wisconsin Conservation Department regarding the use of DDT to solve the problem of Dutch elm disease. An image of a skull and several birds on a tree branch in the rain appeared in the upper right-hand corner.

> Dear Sir,
> It has been found that for the past few years the D.D.T. spraying for controlling the Dutch Elm Disease has caused an increasingly heavy toll of birds. The mortality not only affect [*sic*] the migrant species, that are with us in the spring and fall, but also the familiar summer residents.
> Because of this, a group of interested citizens have made a detailed study of the situation to see if there is a way to reduce the excessive destruction of our bird and animal life.

The group was deeply concerned about the buildup of resistance among harmful insects and the need to increase the quantity and power of the pesticide, and they worried, Larkin wrote, that "the man-made means of control will have to be so lethal as to destroy their master." They hoped to end the use of DDT and the threat to wildlife and humans alike. In another letter, to the chief forester of the Conservation Department, dated June 30, 1957, Marie Thompson, the president of the Animal Protective League, wrote, "We implore you to stop the use of all DDT spraying—which is a complete violation of all of our laws and treaties re: migratory birds, game birds, and wild animals. . . . The death toll from DDT is shocking." Indeed, as early as 1946, the U.S. Fish and Wildlife Service had published a paper that began, "From the beginning of its wartime use as an insecticide, the potency of DDT has been the cause of both enthusiasm and grave concern. . . . DDT, like every other effective insecticide or rodenticide, is really a two-edged sword; the more potent the poison, the more damage it is capable of doing. . . . [It does] not strike the innumerable animal and plant species with equal effectiveness. . . . Certainly such an effective poison will destroy some beneficial insects, fishes, and wildlife." Gaylord Nelson, who was active in creating the Clean Water and Clean Air Acts, was an integral part of the DDT trials that eventually led to legislation that controlled DDT use.

"But after the battle against DDT," Eugene said, his brow furrowed, "there was Sevin." I couldn't remember ever hearing about Sevin, but after some research, I understood Eugene's facial expression. In the mid-1970s, a brand-new miracle insecticide was wreaking havoc on the natural world. Marketed to both farmers and gardeners, Sevin promised to destroy alfalfa caterpillars, armyworms, cucumber beetles, cutworms, flea beetles, green cloverworm, Japanese beetles, lace bugs, leafhoppers, squash bugs, sweet potato weevil, thrips, tomato hornworm, tomato pinworm, tortoise beetles, ticks, ants (except pharaoh, harvester, and carpenter ants), azalea leafminers, blister beetles, brown-tail moths, elm aphids, eriophyid mites, fall armyworms, Fuller rose beetles, plant bugs, scale insects, tent caterpillars, and many more pests listed on the product label. The user was invited to spray fruit trees, vegetables, and flowers, among other things. What the manufacturer failed to mention was that the chemical was deadly to bees—bees who, along with other native pollinators, are responsible for pollinating the very same plants. It was also deadly to humans. In 2004, the EPA issued a statement of assessment for the pesticide.

Carbaryl [or Sevin] is one of the most widely used broad-spectrum insecticides in agriculture, professional turf management and ornamental production, and residential pet, lawn, and garden markets. Although dietary (food and drinking water) exposures are not of concern, carbaryl does pose risks of concern from uses in and around the home. With mitigation measures discussed in the IRED document, carbaryl will fit into its own "risk cup" and will not pose significant aggregate risk concerns. Carbaryl also poses risks of concern to occupational handlers who mix, load, and apply the pesticide in agricultural sites, and to workers who may be exposed upon re-entering treated agricultural areas. Carbaryl poses ecological risks, particularly to honey bees and aquatic invertebrates. With mitigation measures, these occupational and ecological risks also will not be of concern for reregistration.

Carbaryl was the chemical manufactured in the factory in Bhopal, India, where a horrific accident killed twenty-two thousand people and left nearly a hundred thousand chronically ill. The aftermath of this industrial disaster was vividly narrated by Indra Sinha in his novel *Animal's People*. Sinha is a vocal critic of Union Carbide and Dow Chemical.

Both Eugene and Emmett remember the problem with Sevin. They said the chemical companies had worked with universities. "Used to be the kill rate was around thirteen hours. Now it's down to three or four. When they spray at night, by the time the bees get on it, on the corn the next morning, the kill rates are down. Maybe they drag something back, but it's not nearly like it was. Back in '65 we had pictures of just rows of bees dead in front of the colonies. They got hit hard." At that time, consensus again led to changes. Eugene hopes that farmers and beekeepers and chemical companies will come together to solve the new problems bees are facing today. "I'm not against anyone. But everybody has to work together," he says.

These days, Eugene is fighting, along with many other beekeepers worldwide, against the newest in the line of chemicals that are toxic to bees, neonicotinoids. More generally, Eugene combats the effects of harsh chemicals by doing several simple things. He has been moving his bees into smaller farmsteads and locating them near CSA (community-supported agriculture) fields where people practice organic farming, which Eugene feels is helping. But the honeybees face other issues as well, like mites, viruses, and lack of a diverse diet, and the fact that millions of bees are being trucked around the country just exacerbates all of these problems. Eventually, a tired look came over Eugene's face. "Beekeeping has changed," he said, with an atypical sadness in his voice. "It's changed a lot."

We piled back into the truck, threw our long gloves up on the dashboard, and Eugene started the engine. He squinted out at the landscape one last time before putting the truck in gear. A landscape trafficked by his bees. He turned and

19

gave me a lopsided smile. "Okay," he said, "here we go." Glancing at him then, I remembered something he'd said to me the first time we met, as he bent over a hive of bees he'd just cracked early in the spring. "See how pretty they are?" he'd said, and then chuckled to himself. "Beekeeping isn't a job. Beekeeping isn't an art. It's a love," he'd said.

Bouncing along the road toward the honey house, I knew that I needed to learn more about the complex relationships between humans and bees. What was the future of bees going to look like? I had heard that there was a place in China where no pollinators were left, and flowers had to be pollinated by hand. Was that where we were headed? Like Eugene, I felt love around bees. There was a heaviness in my heart. Couldn't we imagine a new, more gentle way of interacting with our community of insects, plants, waters—what Leopold collectively called "the land"? "We abuse land because we regard it as a commodity belonging to us," he'd written. "When we see land as a community to which we belong, we may begin to use it with love and respect." In a world obsessed with profits and computer screens, how might we rekindle or create a sense of belonging with the nonhuman members of our larger community? Especially with something as alien as an insect? How can we create connection and intimacy? Perhaps as children we were more open to that connection.

Child Mind
The Art of Sibylle Peretti

Sibylle Peretti's beautiful work initially interested me for the very superficial reason that honeybees played a role in many of her pieces; she appeared to be another person, like Eugene and myself, who loved bees. I soon discovered, however, that her images were not mere illustrations of children and bees that I would easily forget; these strange, dreamlike images in manipulated glass were like ghosts that haunted me.

But why was I so haunted by Peretti's work? Why did I feel that it so accurately depicted the vulnerability of both bees and humans? Partly, it triggered memories of a very different part of my life and a very different experience with bees.

For a brief few years as a little girl, before my parents divorced, my days were spent floating with my dog between my mother's clay studio, my father's painting studio, and the acres of wildflowers and pine trees that surrounded the simple house my parents had built. The landscape outside the door sang with magic and adventure and early glimpses of the sacred. I tied ribbons on my dog, on days when she was queen of the world, and out we went, marching past

the chickens into the fields to observe what was afoot in her kingdom. The bumblebees, grasshoppers, and wrens paid no mind to us, to the flow of narrative spilling out of my mouth as I reported things to the queen. I was certain that all creatures were sentient, though I didn't know that word at the time. The birds and the bees were obviously living complicated, thoughtful lives. I collected juniper berries like they were jewels and adorned myself with Queen Anne's lace. The wind and the clouds always seemed ready to mirror my emotions. Bliss filled those spaces. From time to time, I lost myself completely in those fields.

After a good walk, the dog and I made our way back to the gravel road that led home. On sunny afternoons after a day of rain, scores of yellow sulfur butterflies collected in the damp places where puddles had been, and I would charge through them, suspended for a moment in a fluttering golden cloud. Other days, a big snake sunning himself in the middle of the road would call upon the knightly part of me, and I would grab a stick and navigate bravely around him. At times, we would seek refuge on the cool piles of

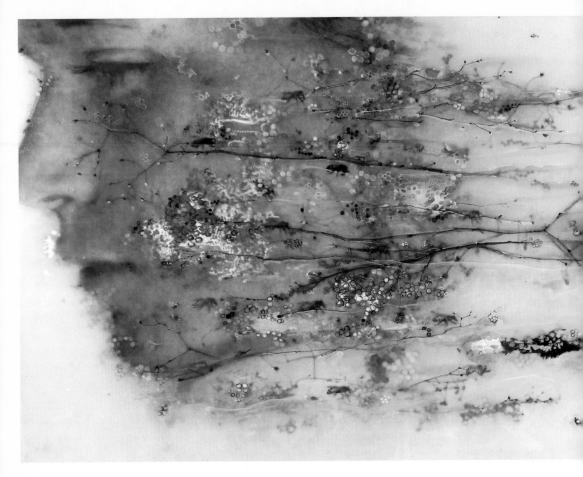

Sibylle Peretti, *Flight*. Glass, 22 × 60 in.

needles under sagging pine boughs where chickadees were conducting their chittery business.

Eventually, returning to the house, I would buzz through my parents' studios to see what interesting things they might be doing there. My mother made salt-glazed cups and bowls for everyday use, and my father made paintings and etchings that seemed like illustrations of dreams to me. If I stopped into the clay studio, my mother often stood up from her potter's wheel, wiped the clay off her hands, cleared a place on her worktable, and gave me some clay to make into something while my dog gratefully took a spot in a shady corner on the cool cement floor. Other times, I'd visit my dad's studio on the other side of the house, and sit down to do a drawing on the floor near his easel or plop down in the rocker to watch him paint.

My father was a man of few words who frequently spoke in metaphor to convey

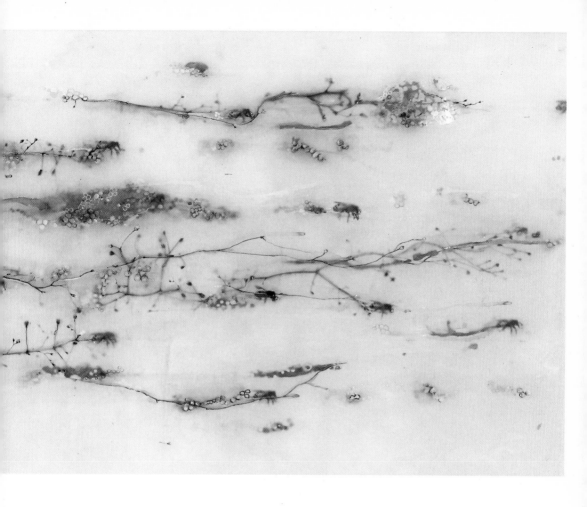

even simple things to me, and who often recited his favorite poems during my visits. Though I didn't understand most of the poems at the time, the rhythm of the language, the soothing sounds of syllables and subtle rhymes lining up, and the beauty of the images completely seasoned my sense of being in the world. Only years later did I learn that one of the poems was called "Directive," by Robert Frost. So many of the lines in that poem became important to me

as an adult: "Back out of all this now too much for us, / Back in a time made simple by the loss / Of detail. . . ."

First there's the children's house of make-believe,
Some shattered dishes underneath a pine,
The playthings in the playhouse of the children.
Weep for what little things could make them glad.

Not until much later would I understand this poem, how at times it was necessary

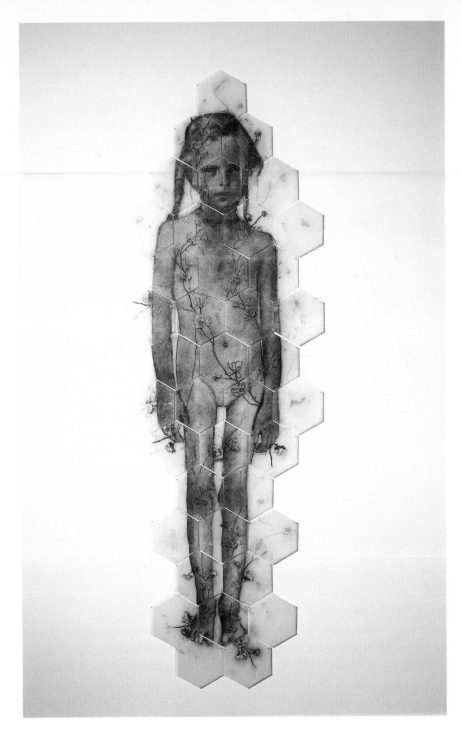

Sibylle Peretti, *Honeychild.* Glass, 100 × 36 in.

Sibylle Peretti, *St. Francis*. Glass, 38 × 23 in.

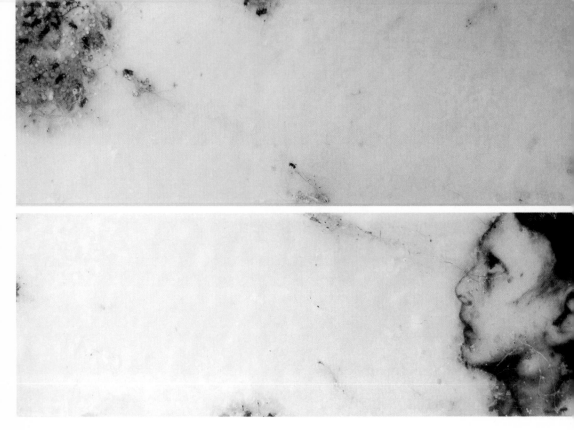

Sibylle Peretti, *The Beekeeper*. Glass, 42 × 60 in.

to relinquish the cynicism and attempt to rekindle a more vulnerable, more wonder-filled space, to remember "what little things could make them glad." And somehow Peretti's work echoes these thoughts.

Peretti, who was born in Cologne, Germany, and was educated by traditional glassmakers there, describes her work in this way:

> I explore the lack of harmony between human beings, nature, and our inability of achieving a unity with the natural world. . . .
>
> While my work hovers between subjects of scientific curiosity, fairy tales, and dreams, I use images of children to open our eyes to a mysterious sensibility we may have lost. . . . I examine the child's identity in a world of adverse layers. The overlay and containment of irreconcilable natures—of disease and beauty, of intimacy and of distance and of innocence and knowledge—have typified the search I have found most important in my work.

Peretti's words articulated beautifully the very ideas I had been mulling over. How there is the desire for unity with the natural world alongside the clear lack of it. How we vacillate always between intimacy and distance.

Peretti's work also articulates the space of creativity, that dreamlike space of curiosity, openness, vulnerability, surrender—the space of the child mind. This state of mind is one I believe we need to remember, to cultivate especially now, as the planet faces so many obstacles. I believe that my movement between the outdoor world of bees

and clover and the indoor world of makers and making allowed me to fuse these two worlds quite naturally and to learn to honor and trust the rich spaces of the uninterrupted imagination. This connection has informed my life. I wrote my first poem to a tree in those more innocent, or perhaps more open, days, and one of my earliest memories is of doing a painting of a flower. But the days of this rather romantic childhood did not last long.

This child-self, like most child-selves, was quickly schooled in the ways of the world: violence, poverty, and abuse (both among humans and against nonhumans), which were a stark contrast to the world I found in the fields and in those studios. Like the children in Peretti's work, the innocent, vulnerable child-self is clouded, distanced, and changed.

Peretti tells a story about how, one afternoon, she came across some old books belonging to her grandfather, a doctor. She found pages and pages of images of children with diseases and rare conditions. What struck her most were the expressions on the faces of the children. She said later that she "felt obligated to transform these children by removing them out of their original realities and pair[ing] them with natural elements. . . . I try to endow them with a higher dignity, which they were deprived of as medical specimens." Disease is a palpable metaphor here. I think many of us recognize a part of ourselves that needs healing. And yet the additions of birds, honeybees, and branches are still somehow separate layers in her artwork. There is a tension at work between the longing for connection

27

between the child and the nonhuman other and the barriers preventing that connection.

I became a beekeeper as an adult after a period of personal difficulty. Being with bees turned out to be a healing experience. I had always been interested in them. I loved watching their small, furry bodies float into an open blossom, where they would nuzzle the fragrant, pollen-covered filaments, never harming the flower but rather helping it multiply. And I loved honey. As I mentioned, my father had taken me to a beekeeper's honey harvest when I was young, and I never forgot the intoxicating sweet smell of that shed. Having the chance to care for bees seemed like a gift. "Beekeepers are chosen by bees," I'd heard someone say. I hoped to be one of the lucky ones.

I rediscovered the privilege of losing myself while watching the bees navigate their intricate collectives—the way they caressed one another's faces, the way they trembled ecstatically as they danced a map to good nectar, the way they worked ceaselessly to make their castle of wax and liquid gold. This was a community based on trust and commitment and what looked to me a lot like love. With bees, I found refuge and wonder.

In her book *The Sense of Wonder,* Rachel Carson wrote about the importance of wonder not only in children but in adults:

A child's world is fresh and new and beautiful, full of wonder and excitement. It is our misfortune that for most of us that clear-eyed vision, that true instinct for what is beautiful and awe-inspiring, is dimmed and even lost before we reach adulthood. If I had influence with the good fairy who is supposed to preside over the christening of all children, I should ask that her gift to each child in the world be a sense of wonder so indestructible that it would last throughout life, as an unfailing antidote against the boredom and disenchantment of later years ... the alienation from the sources of our strength.

I look at Sibylle Peretti's images of children and bees and am filled with an awareness of the precariousness and fragility of the human spirit and the nonhuman beings, all trying to navigate a world of violence and chemicals and disharmonies, but also with a longing for a childlike innocence and openness, a creative space of beautiful connectedness, a place where humans and bees could somehow find even a slightly compromised or constructed harmony. Peretti, to me, is cultivating and reminding us of wonder, something that allows us to envision and create a different world. A sense of wonder, I believe, is absolutely necessary for our survival as a species, and the best path toward connecting with nonhumans and advocating for a world where we can fashion a more sustainable existence for everyone.

It was a rekindled sense of wonder and curiosity, as well as a deep concern about the future of bees, that led me to China to learn about the ways in which humans and nonhumans were working toward remedying the extreme lack of harmony they were facing there.

chapter two

Searching for the Bees of Guangxi and Sichuan

I traveled to China with expectations. In the weeks before my trip, the media had been obsessed with the black air of Beijing, the many thousands of pig carcasses being fished out of a river near Zhejiang province, and accusations about the impurities and additives found in the massive quantities of honey exported by China. I had seen photographic essays of toxic waterways in Shaoxing province and portraits of factory workers in light blue uniforms piecing together plastic dolls for the children of the world—all the terrible side effects of being an industrial giant. I had heard that the Sichuan region had wiped out its pollinators with indiscriminate pesticide use and that the government's solution was to train people to be "human bees." The idea that humans could replace this nonhuman labor force seemed farcical to me. Sichuan seemed a clear example of our dire, beeless, poisoned future. But I wanted to go and see for myself. I wanted to see if there were any bees left and to visit some agricultural regions, so here I was, on the verge of discovery.

My expectations were not all so dark, though. China was home to scroll paintings and Li Po, misty mountain temples and panda bears. And China has a rich history with bees. The Chinese have been using honey in traditional medicine since at least 220 B.C.E. It is said that the Tibetan Buddhist goddess of travel, Chammo Lam Lha, rides a queen bee, and a monk told me that Tibetan Buddhist monks consider it auspicious if a swarm of bees arrives at their monastery. Although most honey producers in China have much more contemporary beekeeping methods, I had read that the Hani people, who live high in the Himalayas, have the skill and patience to tie a little thread with a feather on one end to the leg of a bee so that they can follow her to her hive, where they then gather honey.

And at this historical moment, China is the largest honey producer in the world. So I knew there were bees somewhere. I would travel first to the town of Yangshuo, Guangxi, in the southwestern part of the country, famous for its karst mountains, winding rivers, and agricultural heritage. Then I would head north to Chengdu, Sichuan, and see what I could learn.

I arrived in Yangshuo at night. The bus bounced into the city center, which was teeming with mopeds, taxis, rickshaws, buses, and people. A giant electronic billboard, about the width of a baseball diamond, pulsed with light. Images of mountains and fireworks flashed on the screen high above me. This was not in the guidebook. Taxi drivers shouted for my attention as I made my way into the plaza. After some bargaining, I climbed into the backseat of a cab and was taken down a bumpy road to my abode in Baisha Village, the Yangshuo Tea Cozy.

In the morning, I woke to a chorus of frogs and roosters. The velvet air smelled like orange blossoms. As the sky lightened to a deep blue, the tall thin mountains took shape, like giant monks in robes circling the valley. A bent old man in a straw hat moved about in the nearest rice paddy with a flashlight and a net. Catching frogs, I imagined. When the sun had risen, people on bikes carrying everything from lumber to bunches of chard began pedaling past. I decided to rent a bike and explore.

The valley was made up of several small villages. I rode past one-room stone houses with cement floors, many bearing sacred red banners hung by the door, and other buildings in various stages of construction. Chickens strode through the streets, and small children ate rice out of bowls on front stoops as elders sat nearby drinking tea. What struck me was that almost every bit of green space was used for growing things. Rice paddies and small groves of kumquats and oranges were flanked by gardens full of beans, peppers, and squash. Snap peas curled around bamboo fences, and their sweet white blossoms were covered with bees. Even four-by-four-foot plots between houses were filled with lettuce and kale. The yellow blossoms of rapeseed blanketed the low hillsides. They were also covered with bees. Water buffalo patiently plowed new ground. A woman with her cow and calf stood together in the shade, and when the calf began to jump about, the woman calmly took a small stick and scratched his back and neck until he went back to nibbling grass. A cluster of grayed wooden bee boxes nestled on a field's edge, and bees flew lazily in and out, their legs covered with pollen.

I thought of the many miles of cornfields in my region of the States. I thought of the sterile spaces between the rows, the lack of insects and frogs. I thought of the computerized combines grinding through them. Monocrops radically shift the biodiversity of any agricultural area. This shift is illustrated beautifully in a poignant project by the photographer David Liittschwager. The project involved traveling about the world and dropping metal frames one cubic foot square into different landscapes. Liittschwager would observe these tiny areas for twenty-four hours and record all of the species of plant and insect and animal that existed within, or traveled through, the space during that time. The frame, when perched on the branch of a strangler fig in Monteverde, Costa Rica, was inhabited or visited by more than 150 insect, animal, and plant species. In a garden in Cape Town, South Africa, Liittschwager observed thirty different kinds of plants and approximately seventy kinds of insects. But when he placed his frame in a cornfield in Iowa, he was unsettled immediately by the silence. There was no buzzing or clicking or any evidence of insect life. After twenty-four hours, what he did find was corn, a single mite, an ant, several grasshoppers, and a tiny mushroom "the size of an apple seed." This is a snapshot of what industrial agriculture is doing to our planet: completely devastating the rich biodiversity of insect and plant life.

I knew, in a way, that I was romanticizing, but wasn't Yangshuo's reality much better and healthier than the one I had come from? The diversity of plants. The people and animals connected. The healthy bees. Maybe the villages outside Yangshuo were unique in China. Would the commercialization of the city and the development of this valley eliminate these things? Or would these farmers become the models for a better world? The Li River drew more tourists all the time. And I had noticed that some of the local farmers had begun wrapping their papaws (*Carica papayas*) in plastic. There was even a large royal blue building that I was told was a future vegetable-packaging plant. This is called progress.

Before I left, I decided to look up some of the artists who lived in the region. Winding through the stone streets of an ancient city near the Li River, I came upon a shop festooned with red lanterns and a sign that announced "Peng Family Fan Painting Technique Factory." Beside the open doors of the shop hung several newly glued silk fans, splayed open like clamshells, drying in the sun. Inside, scroll paintings and fans covered every inch of wall space, while others dangled from the ceiling. In one corner of the shop was a table covered with paints and brushes; in another was silk in various stages of readiness. Mrs. Peng, the proprietor, had been trained to paint and handle the silk by her grandparents, who were still active painters. Stylized watercolor paintings of flower blossoms, pandas, pairs of koi, tigers, insects, and mist-covered karst mountains mirrored back to me my romanticized vision of this region. Humans, if they appeared at all, were dwarfed by huge stone cliffs and broad rivers. Nowhere were there images of taxis or electronic billboards. Instead, the images all seemed to affirm the harmonious landscape where the well-known poet of the Tang dynasty Li Po found peace. Sam Hamill translates Po's poem "Zazen on Ching-t'ing Mountain" like this:

The birds have vanished into the sky.
Now the last cloud drains away.
We sit together, the mountain and me,
until only the mountain remains.

It seems that the Peng family painters were painting the images of the landscape that many still hope to dissolve into—a landscape in which the human is still less powerful, which is somehow comforting.

Images like these offer a stark contrast to the photographs of other parts of China by Edward Burtynsky. In many of the images in Burtynsky's collection *Manufactured Landscapes*, the human is once again diminished. Instead, his work evokes a sense of what many have called the "industrial sublime." I'm thinking of one photograph in particular of Three Gorges Dam on the Yangtze River. The viewer of this photo looks into a vast cement canyon, the very guts of the dam, with giant turbines lining one side and steel cranes thrusting skyward in many directions, like the arms of a giant insect. The constructed landscape all but erases the tiny handful of humans in the photo. Burtynsky seems to want us to notice not only that the natural world has disappeared by our hand, but that we ourselves are also disappearing into the industrial landscape we have made. This certainly describes the story of certain parts of China. The images made by the Peng family tell another story, just as my own experiences depict still another. While I recognized that the Pengs' imagery was designed to sell, I felt that these painters were preserving a vision of the human relationship with land that honored at least the beauty of plants, animals, water, and bees, which I suppose I secretly hoped would inspire their audience to protect it. It is probably telling that I left the store with a long scroll painting of the misty mountains that had a tiny figure at the bottom of a person walking a cow along the banks of a river.

But what would Sichuan be like? Sichuan could not possibly be like this, could it? I imagined a bleak scene with no bees or other pollinators, little crop diversity, and the hand pollinators exhausted from climbing up and down in the trees to brush each blossom. There was at least one place where bees were making it, however, not far from Chengdu, Sichuan. It was a Conservation International site where beekeeping was part of a plan to save the panda. This is where I was headed next.

Two days later, I flew up to Chengdu, which is built on the fertile Chengdu Plain, known as "the land of abundance." Chengdu was certainly abundant in people, anyway—fourteen million of them, actually, in the city proper. At the center of Tianfu Square, people poured from the mouth of a brand-new subway system to shop and gape. A twenty-foot-tall woman in Calvin Klein underwear danced in slow motion on the side of a giant building across from a popular Starbucks. Chengdu's young professionals paraded past in their neat suits, eyes glued to their iPods and Droids. Mopeds and rickshaws competed with cars and pedestrians for the right-of-way. Away from the city center, smaller shops dominated: flower shops offering jars full of bright yellow, purple, and magenta blossoms; meat shops draped with strings dangling headless chickens; broom and mop shops; and Chinese medicine shops crammed with baskets of dried fish lungs, ginseng root, ginkgo, dehydrated deer fetuses, and small jars of royal jelly and honey. This was a real city.

My new address was in the ancient part of Chengdu, near Wenshu Monastery. The maze of narrow stone streets zigzagging through the Tang dynasty–style buildings was filled with tiny stalls selling Buddhist prayer beads, hand-painted silk fans like the ones I'd just seen being made, and panda knickknacks. Chengdu prides itself on being the home of the giant panda, China's national treasure. Small shops and giant malls alike sell all types of panda paraphernalia: panda bracelets, panda paintings, panda T-shirts, panda key chains, panda chopsticks, panda hats, panda backpacks, and, of course, stuffed pandas. The bamboo forests on the nearby mountains were once home to a flourishing panda population. As with many species all over the globe, development and deforestation have eliminated much of the panda's natural habitat. In response, the government had been working to save the animal. By most estimates, there are approximately eighteen hundred pandas left in the wild. One approach to saving them is something called "payment for ecosystem services," or PES. My friend Nathan Schulfer, a University of Wisconsin conservation biologist, and his colleague Ming Hua introduced me to one PES project they are working on. The project, sponsored jointly by Conservation International (CI), the Chinese Forest Service, and the regional government, has a conservation site located outside Yingjing, Sichuan, in the middle of a panda corridor. The project encourages villagers to keep bees and harvest honey for money rather than cut down bamboo, thus preserving habitat for the surviving wild pandas. CI had agreed to pick me up and take me out to meet the beekeepers.

Terraced fields and forests made up of ginkgo, bamboo, and flowering crabapple rose up along the road to Yingjing. I wondered whether the Grain for Green program, which aimed to reforest many eroded areas of China, had affected this region. More often than not, the landscape felt quite rural, but occasionally the evidence of industry was present. Suddenly, an almost unbearable chemical smell would overwhelm me as we drove through a portion of the plain covered with the

33

accoutrements of production. Factories billowed smoke, and lots the size of city blocks were filled with shipping containers and crates. But then we would drive back into farmland. Overall, things were very green.

I arrived in Yingjing and was met by the young and vibrant Crystal Zhang of Conservation International, the quiet and friendly Chen Farbing, president of the Yingjing Honey Association, a very solemn official from the forest service whose name I never learned, and a few other CI workers. We piled into SUVs and drove through the town of Yingjing to the honey-production facility. We drove up to a nondescript cement building and made our way to the second floor. Chen Farbing led us into an immaculate series of rooms with bright white tile floors, filled with state-of-the-art stainless steel honey-processing equipment. One room housed a large extractor, similar to the one Eugene had in his honey house, which spun frames full of uncapped honeycomb inside a large drum. The honey was collected and taken to another machine, which heated the honey to 50 degrees Celsius (122 degrees Fahrenheit)—to prevent excessive crystallization, he said. Another room was set up like a lab, with microscopes for examining honey for beneficial pollen grains and for examining the guts of bees for nosema and other diseases. Yet another room was filled with shelves full of empty glass jars, lids, and labels. Chen said that the facility handled the honey of 308 beekeepers (including the beekeepers in the panda corridor), and that they processed three to eight tons of honey a year. A large portion of that honey was purchased by Marriott for their "nobility of nature" program and was served at breakfast in many Marriott hotels around the world. The rest was sold in their store in downtown Yingjing. It was a very impressive operation, and I understood, given the recent accusations about bad Chinese honey, why they wanted me to see the very clean and efficient production end of things. But, to be honest, I was really eager to see some bees. We made our way back out to the vehicles, climbed in, and drove to the edge of the town, where we would meet the beekeepers.

The beekeeper Yin Xianlin's short-cropped silver hair, composed angular face, and dark blue Mao suit gave him a distinguished and rather solemn air. When he started talking about his bees and discovered that I too was a beekeeper, however, his face grew radiant, and he led me quickly to meet his bees. As he slowly opened his thriving hives—without a veil, I might add—the smell of wax and propolis filled the air, and I suddenly felt right at home. I marveled at the seething mass of lovely insects he proudly held up to me, one frame at a time. I found myself sighing loudly, and nodding to everything he pointed out. He pointed out the queen, and we marveled over the many shades of pollen packed tightly into cells in one frame, the cells filled with honey in another frame, and the healthy brood in yet another. I felt, for perhaps the first time in my life, what I think musicians sometimes feel with other musicians, even of different cultures and language—a kind of connection, a harmony, that occurs among people who deeply love the same thing, and can therefore share and communicate inexplicably, and without any recognizable language at all, but rather with sounds and gestures.

He explained, through Crystal, that these were not Italian but Asian bees, which were smaller and gentler, in his opinion, and had lower populations than Italian colonies. His bees fed on vegetable blossoms, but later in the season he took them to savor the mountain flowers of the golden cypress, the Chinese gall, and the camellia. I asked whether his bees were experiencing problems with mites or anything else. He prevented mites, he said, with traditional Chinese medicine. Once he had closed up the hives, he walked me out to his garden and showed me the specific plants that he used to prevent mites. I couldn't identify them. When I asked about annual losses of hives here in this valley, he and Chen looked confused. I explained that in the United States this year, beekeepers had lost on average 35 percent of their hives. They shook their heads: no, no, we don't have anything like that. But chemicals, I asked, do you have trouble with them? Oh, yes, they knew there could be trouble with chemicals, but that wasn't an issue here in Yingjing. They kept their bees away from pesticides. I was puzzled. This was Sichuan. Didn't fruit farmers have to pollinate by hand? In fact, I was told, only two very small areas had experienced this problem and had opted for this solution, but these areas were clearly the exception.

In two small counties in Sichuan, one of them neighboring Yingjing, the use of pesticides had so devastated bee populations that all the pears and apples were pollinated by hand. Land reform in the mid-1950s enforced the formation of rural farm collectives. The counties of Maoxian and Hanguan were planted heavily with pear and apple trees, approximately 280,000 in all. In the 1980s, another round of reforms, called the household responsibility system, returned production responsibilities and benefits to individual families. Because the plot sizes were small (approximately one-third of an acre per family), people wanted maximum yield. Until the mid-1980s, no hand pollination was used because insect pollination was sufficient. The locals say that in the early 1980s, a plague of fruit lice required heavy doses of pesticides. That is when pesticides began to be used regularly, a practice that continues today. As recently as 2012, farmers in these regions were applying pesticides, herbicides, and fungicides, on average, seven or eight times a year. The heavy pesticide use eliminated the honeybees and other native pollinators. The villagers were forced to come up with a pollination strategy. Their solution was pollinating by hand.

Pears and apples require the cross-pollination of more than one cultivar. When there were still insects in the area, orchards needed at least 20 percent of the trees to be what are called "pollinizers." The most popular pears came from the Jinhuali tree, which were best pollinated by the Yali tree. Once the communities switched to hand pollination, they could plant fewer Yali pollinizers and more fruit-bearing Jinhuali trees because they could control the pollination themselves. Over time, they learned to read the flowers and know exactly when they were ready to "mate." Apparently, women are considered superior to men at this sensual, delicate practice, though many men do it, and children are not allowed to participate at all.

In order to pollinate properly, you must pick the blossom just as the stamen is full and ready to release its pollen. When the blossom takes the shape of a bell, it is ready. The blossoms are plucked quickly and carefully by hand and taken to people's homes, where they are dried in boxes covered with cloth, or sometimes under lamps. At just the right moment, the human pollinator holds the stamen in her hand and gently rubs a tiny brush along its sides until the stamen bursts and the golden pollen is released. The pollen is collected, bottled, and kept warm until it's time to take it out to the orchards again. Timing is everything. Pollination will succeed only if the female flower is ready. A few lines from a poem called "A Blossom Tree" by Xi Murong come to mind, about a woman whom Buddha turns into a tree:

> Blossoming, discreetly, under the sun
> Every flower is my previous life's yearning.
> When you trek near, listen carefully:
> The trembling leaves are my longing . . .

Luckily, the Jinhuali generally opens its blossoms about a week after the Yali. The pollinator prepares her brushes most commonly by inserting the end of a chicken feather into a cigarette filter. Many brushes are needed for the brief season, as the trees are pollinated more than once. The pollinators know the Jinhuali is ready to be fertilized when the edges of the stigma begin to darken. At this moment, the pollinator climbs into the tree, dips her brush into the pollen, and lightly brushes the stigma's opening. Because the flowers open on different days, and because they repeat the process, it takes more than a week to pollinate the orchard. A very efficient person can pollinate about five hundred blossoms a day. One bee can pollinate up to five thousand blossoms a day. One normal hive, with a population of approximately forty thousand bees, can easily take care of an orchard in a couple of afternoons. The window for this pollination practice is obviously very small, and unfortunately I had missed it by a few days.

Photographs of this practice show up in many contexts—in lectures by Dr. Marla Spivak about Colony Collapse Disorder, in films such as *Queen of the Sun, More Than Honey,* and *Vanishing of the Bees*, and in academic papers—all signifying the urgency to wake up and shift our practices to avoid a beeless world where humans would be responsible for pollination.

Sichuan University professor Tang Ya, a thin, bespectacled intellectual, has researched these regions extensively. He met me one day on campus with a huge smile and led me to his large office. His work for the International Centre for Integrated Mountain Development, based in Nepal, had led him to examine areas in several countries, including Burma, India, China, Bhutan, and Bangladesh, that had suffered large-scale pollinator losses. Only in China had the solution been to use human bees. According to Professor Ya, there are certainly bees in these regions in Sichuan now, but the beekeepers refuse to take their bees anywhere near those orchards. Even when fruit growers offered to pay the beekeepers, the beekeepers would not endanger their bees. Thus the fruit on those hillsides was 100 percent human-pollinated, a practice that was getting too expensive to continue.

Beekeepers were refusing to take their bees into those orchards? This idea was stunning to me. What would happen if beekeepers in the United States and other places simply insisted that farmers change their practices? I knew that some blueberry farmers, like Dennis and Shelly Hartmann of True Blue Farms in Michigan, with the help of Dr. Rufus Isaacs, an entomologist at Michigan State University, were experimenting with integrated pest management and native wildflower plantings on crop edges to promote healthy pollinators and reduce pesticide use by encouraging beneficial, pest-eating insects. If beekeepers united, could they demand more changes like these that could save the honeybees?

I asked Tang Ya what the future of Sichuan would bring. Tang replied that some farmers were planting different fruit crops, which either required fewer pesticides or were self-pollinating, and that this seemed hopeful to him. But things were changing. Younger people were moving to the cities for jobs that made more money, he said. They didn't want to farm or keep bees. And development was changing Sichuan. New factories were being built all the time, taking up land. By the end of our conversation, his face was not so cheerful. He, too, was very worried about the future of bees and beekeeping.

The Conservation International project, according to Crystal and Nathan, was going quite well. They needed to establish some ways to monitor the progress of the ecosystem services site, but they were hopeful. The bees were helping save habitat for the pandas, and the panda habitat provided healthy, organic forage for the bees. And the beekeepers were making good money. But funding was promised for only a few years and would soon run out. The continuation of the project was uncertain.

On my way home, I made a stop in Hong Kong, the third-largest port in the world. The city churns with the energy of global trade, its harbors full of cargo ships stacked with containers, surging constantly in and out. While sipping tea with acacia honey, I caught up on the latest news. I read an article about the twenty thousand pig carcasses found in the Huangpu River. Small farmers north of Shanghai had been accused of dumping animals too sick to be sold for meat; a crackdown on meat quality had created a surplus of dead animals. Because these small farmers had little land, there was no space for proper burials. The river was their answer. In his response to the situation, *Atlantic* magazine journalist Matt Schiavenza wrote, "Can the [Chinese] government make sure this sort of thing doesn't happen again? The agriculture sector is consolidating, as millions of Chinese leave the countryside for the cities each year. Bigger farms will create economies of scale and standardization—leading to cheaper, more reliable pork for Chinese consumers—and will [increase] perverse incentives to dump dead livestock in the river." I thought of the woman scratching the calf's head in Yangshuo. I thought of Yin Xianlin and his garden full of healing plants. Standardized farms, to me, meant CAFOs (confined animal-feeding operations) and miles of genetically engineered corn. They meant systemic pesticides that leach into the soil, the water, and human and bee bodies. Standardization had its merits, but even greater drawbacks, especially in the long term. Vandana Shiva radically suggests that small, diverse food crops are the answer to poverty in India. She believes that food commodities and high yields of single crops are destroying ecosystems and making people sick. Many people scoff at her. That's a return to the old ways, one interviewer told her scornfully in a recent interview. No, it's the way to a sustainable future, Shiva asserted.

Perhaps Yangshuo and Yingjing stand at a threshold. Perhaps large standardized farms will replace the bamboo fences covered with sweet peas at the edges of the orchard of twelve orange trees near the Yulong River. Perhaps monocrops of wheat and soy will be planted, harvested, and sent to Hong Kong and then shipped out into the world. Or perhaps, instead, the world will see these places as exemplary. For now, the farmers, beekeepers, pandas, and bees in Yangshuo and Yingjing have what keeps them all living in what looks to be a fairly harmonious situation: a variety of vegetables and fruits, diverse habitats in the mountains, and low pesticide use. I sipped my tea and stared out at the busy bay.

The Microscopic Sublime
The Art of Rose-Lynn Fisher

I was still mulling over my experiences in China and how they had radically altered my understanding of the dynamic, interconnected relationships between bees, flowers, animals, and beekeepers, and also my own part of the world, when I first became interested in the work of Rose-Lynn Fisher. The experience of seeing Fisher's photographs radically changed the way I thought about the honeybee herself. While Edward Burtynsky's images of the industrial sublime in China invite the viewer to experience a shift in thinking about humans and the built environment, Fisher's images invite a needed shift in the human perspective on bees.

Fisher is a photographer living in Los Angeles. She received her BFA from the Otis College of Art and Design in Los Angeles and also studied at UCLA and UCSC. Her work, which she describes as exploring "a sense of place, where distance may be measured in miles or magnifications," has been featured in galleries and museums all over the world and has been reproduced in periodicals like *Smithsonian*, *Harper's*, *Time*, the *Guardian*, and many others. She has been using microscopes to make images

since the early 1990s. The pieces included here are from her book *Bee*, published in 2010 by Princeton Architectural Press.

The first time I saw Rose-Lynn Fisher's bee series, I could not stop staring at the photo called *Eye*. The hexagonal patterns on the surface of the honeybee's eye, I suddenly realized, were a mirror of the design of the honeycomb bees built inside the hive! Could this be the reason why they designed their homes as they did? Did their vision literally define their architecture? I was mesmerized. It turns out that this same thought had occurred to Fisher when she first saw the pattern. Perhaps it is simply the most efficient way to deal with circles, she thought. Perhaps she's right. Fisher's images elicited all kinds of questions and made me see that the bee was much more complicated than I ever could have imagined. Her images force a kind of intimacy and respect that I was unprepared for. And, to be honest, it was a little terrifying. The proboscis, especially, no longer looked delicate, but rather like a big hairy arm. In addition to the tiny hexagons of the surface of the eye, there were also hairs growing out of it,

Rose-Lynn Fisher, *Bee*. Originally published in Rose-Lynn Fisher, *Bee* (New York: Princeton Architectural Press, 2010).

which looked more like strange trees in a Dr. Seuss landscape than anything remotely like an eye. The honeybee was not what I would call beautiful at this magnification. But in imagining what these features of the insect could mean, I was overwhelmed. The antennae, for example: if each of those tiny hairs could pick up sensation the way hairs on my own arm did, what a remarkable sensory experience the bee must have in touching another bee or the petal of a flower! My perspective was dramatically altered, and I could not think of the bee in quite the same way as before. It occurred to me that the images revealed a microscopic sublime. If so, was this useful in some way? Could it help people think differently about little beings like bees?

The romantic sublime is traditionally associated with enormous powerful things like mountains and intense storms. "The passion caused by the great and sublime in nature . . . is astonishment; and astonishment is that state of the soul, in which all its motions are suspended, with some degree of horror. In this case the mind is so entirely filled with its object, that it cannot entertain any other," said Edmund Burke. In his *Philosophical Enquiry into the Origin of Our Ideas of the Sublime and the Beautiful* (1757), Burke discussed at length his distinction between the sublime and the beautiful, claiming that the sublime was associated with "the infinite . . . and terror." The sublime of the romantic period also became associated "with powerful emotions, with spiritual and religious awe, with vastness and immensity, with natural order in its grander manifestations and with the concept of genius."

Artists like Caspar David Friedrich and poets like William Wordsworth and Percy Bysshe Shelley tried to evoke the vast and mysterious powers of nature in their work. People took grand tours through Europe to see awe-inspiring, edifyingly sublime landscapes like "the Alps, abysses, forests, mountain ravines and torrents." But what happens when we diminish sublime nature? Can the miniature or typically invisible also produce this kind of inspirational awe? And what do we do with that feeling? Does a sense of responsibility follow?

In November 1783, the Montgolfier brothers successfully launched the first manned hot-air balloon. As Richard Holmes has suggested in his book *The Age of Wonder: How the Romantic Generation Discovered the Beauty and Terror of Science,* "The dream of flight had haunted men— especially poets, satirical writers, and impractical fantasists—since the myth of Icarus. European literature was full of unlikely bird machines, flapping chariots, flying horses and aerial galleons. None of them was remotely practicable." After 1783, however, humans had taken to the air. This new phenomenal ability pushed up against the notions of the sublime in nature. Suddenly, all that had seemed gigantic and insurmountable was miniaturized in the gaze of the balloonist. "Ballooning produced a new, and wholly unexpected, vision of the earth," Holmes writes. "It had been imagined that it would reveal the secrets of the heavens above, but in fact it showed the secrets of the world beneath. The early aeronauts suddenly saw the earth as a giant organism, mysteriously patterned

Rose-Lynn Fisher, *Eye* (above) and *Antennae* (opposite). Originally published in
Rose-Lynn Fisher, *Bee* (New York: Princeton Architectural Press, 2010).

and unfolding, like a living creature. For the first time the impact of man on nature was clearly revealed: the ever-expanding relationship of towns to countryside, roads to rivers, cultivated fields to forests, and the development of industry." The earth was suddenly a small, living planet, "a giant organism" that we all shared, although this awareness did not seem to initiate any real sense of stewardship. If humans could surpass the sublime, shrinking the very mountains, where could we find awe and astonishment?

In the first hot-air balloons, humans were able to sit in positions of power over the things that had previously inspired awe. The ability to look down on a mountain was the first step to raise us "above nature." Eventually, in the 1970s, technology allowed us to see the entire earth from space, a blue and white swirling marble in the center of a dark universe, giving us simultaneously the feelings of fragility and power. Today, satellites feed us constant images of our cities, our roads, our homes, and bring them to our own computers through Google Earth.

The impact of human technology on the planet has been so profound, in fact, that Paul Crutzen, Eugene Stoermer, and many others agree that our current epoch must be called the Anthropocene. Our impact on climate alone since the Industrial Revolution is now shifting weather systems and landscapes in ways that we have only begun to understand. This current notion of the Anthropocene—the idea that our impact is so great that we have altered the planet geologically, controlling species, curating

wilderness, changing our atmosphere—has led writers like Ronald Bailey to suggest optimistically that, "over time, we will only get better at being the guardian gods of Earth." Such a thought only perpetuates the idea that human beings are capable not simply of controlling the earth but of doing it well. Rob Nixon responds, "But for others, talk of *Homo sapiens* as god species, as Earth's surrogate divinity, is positively chilling. Hasn't a hubristic mindset of earth mastery, of dominion over nature, gotten us into this mess as unwitting geological actors? Earth mastery, moreover, conjures up disturbing associations with the race, gender, and class hierarchies of the selective enlightenment." And in truth, we are not in control, regardless of our destructive impact.

So where is the sublime, and the feeling of awe it inspires, to be found in a moment of monumental human impact on the planet? And is it important? Many of us may no longer truly feel less powerful than a mountain. Humans can remove the tops of mountains to extract coal. Humans can dam torrents and redirect rivers. We can split the atom itself. As we have seen, artists like Edward Burtynsky are hoping that images of the industrial sublime will raise awareness of the destructive nature of our current practices. The Australian painter Mandy Martin makes enormous images of mines and devastated landscapes that she hopes will awaken her viewers to the damage of our overconsumption of resources. These new images of our massive impact absolutely arrest and move the viewer, but they may also reinscribe a sense of human power.

Rose-Lynn Fisher, *Wing*. Originally published in Rose-Lynn Fisher, *Bee*
(New York: Princeton Architectural Press, 2010).

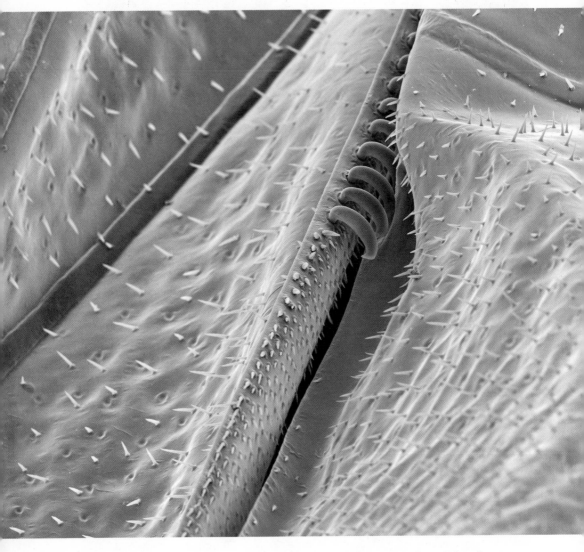

Rose-Lynn Fisher, *Wing Seam*. Originally published in Rose-Lynn Fisher, *Bee* (New York: Princeton Architectural Press, 2010).

But what about the experience of seeing microscopic realities? With the invention of the electron microscope in 1931, humans gained the ability to magnify things that had previously been invisible to us: plant cells, blood cells, viruses. The electron microscope used electrons rather than light to illuminate the examined object, which radically magnified the potential for seeing. I imagine that looking at a bee under a microscope for the first time must have been disconcerting, as seeing is often described by the newly sighted. In her beautiful essay "Seeing," Annie Dillard describes how people who had undergone surgery to heal their blindness had difficulty, at first, making sense of what they were seeing. Lacking an understanding of depth of field, for example, they found it hard to perceive where one shape ended and another began. It seems likely that one would need to learn how to see magnification in the same way. How could you know whether you were looking at a pollen grain, or an antenna, or a leg, when they look so radically different at the new size?

In her book *On Longing,* Susan Stewart meditates on the effects and fascinations of the miniature and the gigantic. "Whereas the miniature represents closure, interiority, the domestic, and the overly cultural," she writes, "the gigantic represents infinity, exteriority, the public, and the overly natural. . . . The miniature offers us transcendent vision which is known only through the visual. . . . Our most fundamental relation to the gigantic is articulated in our relation to landscape."

What happens, then, when the microscopic becomes the gigantic, when the image of a bee's wing is magnified to the point that it takes on the qualities of landscape? Do these binaries collapse? Does the firm boundary between interior and exterior become blurred, or melt away altogether? Do we lose our sense of self and other? Do we begin to wrestle with the question of what is infinite and what has closure? Rather than sense the transcendence we may feel in looking at a tiny insect in our hands, do we begin to understand the unfathomably complex reality that exists outside our ordinary sensory experience? Could we, perhaps, on seeing work like Rose-Lynn Fisher's magnified bees, begin to reacquaint ourselves with awe? With humility? Even with wonder and astonishment, as we contemplate the marvelous complexity that exists outside our control, not made by us?

Robert Frost's poem "A Considerable Speck" describes how his perspective shifted when he recognized suddenly that a speck on his stack of paper was actually a living being.

Plainly with an intelligence I dealt.
It seemed too tiny to have room for feet,
Yet must have had a set of them complete
To express how much it didn't want to die.
It ran with terror and with cunning crept.
It faltered: I could see it hesitate;
Then in the middle of the open sheet
Cower down in desperation to accept
Whatever I accorded it of fate. . . .
Since it was nothing I knew evil of
I let it lie there till I hope it slept.

The narrator of Frost's poem suddenly recognizes the intelligence of a nearly invisible creature and is moved to let it live; the poem offers this awareness to the reader. Like Frost's poem, Fisher's images create awareness and make the invisible visible. She does not want us to ignore the magnificent complexity of bees. She invites us to see their power. If Wordsworth had seen these images, would he also have been overwhelmed by their power? Could he have said, as he did in "Lines Composed a Few Miles Above Tintern Abbey":

While with an eye made quiet by the power
Of harmony …
We see into the life of things.

The power of a honeybee is certainly unlike the power of a thunderstorm or a tsunami or a dam, but the honeybee has the power to pollinate entire crops, has scent glands powerful enough to detect diseases and bombs, is capable of the alchemy that is honey making, and has a formidable sting. These are obviously powers that affect humans and that we must recognize and respect.

Rose-Lynn Fisher's images confront us with a different perspective on these imperiled insects whose lives are completely enmeshed with our own. Hopefully, this perspective shift that Fisher's images offer, this exercise in looking more deeply, may encourage a deepened awareness and a shift in our behavior toward those insects. In southern Africa, I discovered some innovative projects that depend on the power of the honeybee and the complexity of interspecies relationship, projects that were helping not only the honeybee but humans and elephants as well.

Bullroarers, Elephants, and Honeybees

What do elephants have to do with honeybees? you might ask. I certainly had never come across any meaningful connection between them, but after meeting people in China who were keeping bees in part to protect pandas, I was on the lookout for other surprising relationships. In southern Africa, I discovered a wide range of intricate relationships that emphasized that there is no single answer to the bee problem—or to any ecological problem that we face, for that matter—but rather a complex mosaic of relationships that need to be considered and balanced. Elephants and honeybees are two species in one such mosaic. This mosaic is made possible by the fact that elephants are afraid of bees. This fear may be one part of a solution for a problem that humans, elephants, and bees are facing: the problem of competition over scarce resources.

Global water shortages are creating difficulties for humans and other animals worldwide. Migratory animals depend upon water holes to survive, and many of those water holes are disappearing. The elephant, a keystone species in the ecology of southern Africa, has been forced to travel farther in search of water and food. The elephant population throughout most of Africa has seen dramatic reductions in recent years, due to both the ivory trade and loss of territory, yet elephants also maintain celebrity status given their role in African tourism. In a 2010 film made by Moritz Zimmermann for the nonprofit organization Save the Elephants, Iain Douglas-Hamilton says, "The future of all wildlife is bound up with our own species. We at the moment constitute the greatest threat and the greatest hope for their survival." Douglas-Hamilton, the founder and CEO of Save the Elephants, has spent a lifetime working to help elephants survive in a world dominated by humans. Not only has he accomplished groundbreaking research on elephant behavior, drawn worldwide attention to the problem of poaching, and installed GPS tracking systems for monitoring elephant movement, but he has also devised an innovative solution for one aspect of human-elephant conflict: bees.

Very often, villages and agricultural areas lie in the path of migrating elephants. An elephant looking for food or water might venture into a village or a garden or a farm and eat the valuable crops and destroy property. As farms expand and water sources dry up, this occurs more often. Most fences are not strong enough to keep elephants out. While electric fences seem to work, high costs and the lack of electricity in some areas make these fences unfeasible for most villages. Scare tactics employed by the villagers often lead the elephants to charge, causing more damage and sometimes even injuring humans. The frustrated villagers sometimes resort to shooting at the elephants and, occasionally, killing them.

In 2002, Iain Douglas-Hamilton and Fritz Vollrath came up with the idea that African bees might be useful in promoting nonviolent elephant-human interactions, based on the fact that elephants are afraid of being stung by a bee. While an elephant's skin may seem rough and tough, elephants are, in fact, extremely sensitive creatures. The skin on an elephant's belly and behind her ears is much thinner than that on her back. In these areas, she is susceptible to tick bites and bee stings. The eyes and inner trunk are also very sensitive to bee stings. Douglas-Hamilton and Vollrath wrote an article in which they described an elephant who was attacked by a group of bees, and his eyes grew so swollen that he was blinded for twenty-four hours and only recovered after receiving large doses of antihistamines. The elephant's trunk is loaded with nerve endings, which give it a keen sense of smell. A bee sting on the inner trunk is terribly painful, and elephants will go out of their way to avoid it. Knowing this, Vollrath and Douglas-Hamilton designed an experiment in which they put beehives into trees that would normally be tasty to elephants. The elephants left these trees alone.

In 2007, the biologist Lucy King began building on the work of Vollrath and Douglas-Hamilton. King further explored the relationship between elephants and African bees, and soon began a collaboration with farmers in Kenya, Tanzania, and Mozambique. King and Save the Elephants have been building "beehive fences" ever since. A line of beehives dangling between large fence posts serves as an excellent deterrent to hungry elephants. In addition to stopping the crop raiding, the beehives produce honey, which village beekeepers can sell or eat. The bees pollinate the crops and have plenty of forage for themselves. The elephants avoid the fences and stay within elephant-friendly corridors. The humans and the elephants are no longer in direct competition, and the bees have enough to eat.

This use of bees to save elephants is another fine example of multispecies collaboration. The project has been so successful that it has been expanded into many areas in southern Africa, including the Makgadikgadi region of Botswana, home to the San people.

The human-bee relationship in southern Africa is certainly not a new one. In fact, humans were in relationship with bees long before they were used for pollination. The San people of southern Africa recorded their experiences of honey gathering during the Stone Age. Some of the simple and elegant paintings on the shelter walls of Ndedema Gorge and elsewhere in the Drakensberg Mountains clearly illustrate the honey hunter: a person, surrounded by bees, perched precariously on a ladder beneath a hive built on the underside of a rock outcropping. Evidence of the importance of bees can also be found in their poems and prayers.

San Honey Hunter's Prayer
I am weak from thirst and hunger.
  Abo Itse, let me live . . .
Let me find sweet roots and honey,
  let me come upon a pool.
Let me eat and drink. Ho Itse,
  give me that which I must have.

Honey was an important food source, especially because during times of crop failure, honey was often still available. Not only was it consumed as food; honey was also used in rituals because of its role in San mythology. Harald Pager describes its mythological significance: "Particular magico-religious importance was also accredited by the Bushmen to bees and honey. Honey features as a magical potion in some of their myths of genesis, the gods and spirits are thought to be fond of honey, while the great god of the !Kung Bushmen is the protector of the bees and his wife is the 'mother of the bees.'"

San methods of finding honey were uniquely multispecies projects. The San partnership with the greater honeyguide, a tiny bird, benefited both bird and human. To support its diet of honey, this clever bird relies upon a large mammal, sometimes a human, because the bird cannot open the hive on its own. Upon finding a hive, the bird uses her song to attract the attention of potential partners, and then leads them back to the hive. Honey hunters recognized this habit of the honeyguide and learned to whistle to the bird to keep her interest, and then followed her to the beehive, where both human and bird could feast on honey.

While the practice was rather destructive to the bees, this communication with the bird brought the riches of the bees to human tongues.

But the San didn't stop at whistling for the honeyguide. They wanted to communicate directly with the bees while the bees were swarming. Capturing a swarm and depositing it in a hollow log gave the San more control over the honey supply. While their methods differed from the practices of beekeepers today, the motive was the same. House the bees in a place with easy human access and harvest as much honey as you please. Today, buying packages of bees is the most common way to get resident bees for your hives, but many beekeepers in the United States still capture swarming bees. Bees swarm when their population grows too large for their space. The queen bee ventures out with a little more than half the worker bees from the original hive in search of a new place to live. The bees left behind raise a new queen. While the scout bees in the swarm are out looking for prime real estate, most of the others cling to a branch or other armature in a large clump that resembles the shape of a bear's head. The queen bee is nestled safely in the center of the clump. Bees have no territory to defend during this move, and so they are more docile than usual. Beekeepers take advantage of this moment to trap them and take them back to their own hive boxes. Typically, a beekeeper merely has to carefully clip the branch, lower it into a box, and then shake the bees into an awaiting hive body.

The San employ a different method of swarm catching, "calling the bees down" with a bullroarer. A bullroarer is a thin piece of wood on the end of a long cord that is spun around in great circles in the air. The resulting sound is a haunting, guttural humming, similar to that of the Australian didgeridoo. The San believe that the bullroarer beckons bees and can even cause a swarm. Celebration and dancing follow the capture of a swarm. A San person explained that "people used a bullroarer to cause the bees to swarm and make honey which they collected in leather bags and took home to the women. When they had satisfied their hunger, the band danced all night, the women clapping and the men pounding their feet until all were enveloped in a cloud of dust. Today the !Kung still consider bees to be very potent, and they believe that a dance performed at the time when the bees are swarming is particularly effective." I admit that eating lots of honey and dancing all night seems like a ritual we might consider revivifying. But that is not the point. The point is that these remarkable communications between bird and human and between human and bee are examples of how, historically, we have had significant moments of multispecies enmeshment and multispecies communications that require intent listening, observation, and seeing. I found the activities of the San still resonating today in the work of South African artist

Kim Gurney, but her work pushes us to consider our contemporary relationship with bees as well. Her installation in an exhibition called *Ecotopian States* featured a sound loop of the bullroarer. The gallery following this chapter features her thought-provoking work, inviting us to revisit history and rethink the present moment.

While the San developed systems of communication with the bird and the bees, they also, frankly, were simply stealing honey from the bees in a very clever and perhaps reverent manner. While our interdependence with honeybees and flowers continues, our once poetic reverence for those relationships seems to have suffered in the competitive fever of industry and capitalism. Contemporary industrial apiculture is much more destructive to bees than the San ever were.

If we are truly listening to the nonhuman today, the work of scientists and beekeepers can be a valuable guide. How do agricultural practices and other environmental factors affect the bees of southern Africa? A 2003 report called "Crops, Browse, and Pollinators in Africa: An Initial Stock-Taking," produced by the African Pollinators Initiative with support from the UN Food and Agriculture Organization, examined the state of pollinators in South Africa. The report makes clear that because agriculture depends on insect pollination, serious caution must be taken when using pesticides. This was before Colony Collapse Disorder had appeared on the global stage.

At Penn State University, in the United States, entomologists Diane Cox-Foster, Christina Grozinger, and their colleagues have spent several years trying to sort out the causes of CCD and honeybee declines. Their team has examined the vectors of a variety of viruses, including American foulbrood and Israeli acute paralysis virus, and the effects of a wide range of pesticides. Their findings suggest that the ingestion of pesticides and the widespread use of monoculture farming have weakened bees' immune systems. The results of tests for pesticides in hives provide a sense of the ubiquity and diversity of these poisons. "The cocktail of pesticides that bees are exposed to is most striking," according to the Penn State Extension website. "On average, there were six pesticides in each pollen sample, up to 31 pesticides in a single pollen sample, and 39 pesticides in a single wax sample. Think about what the first thing your doctor asks you when you visit the clinic: are you taking any medication? We know that multiple chemicals in the body can react to create toxins. The interactions between these pesticides can be very complicated and difficult to predict." As Sue Hubbell points out in *A Book of Bees*, "Some agricultural pesticides act so rapidly that the bees die in the field, but with others, the bees struggle back and die in convulsions in the hive." As noted above, in 2013 it was established that neonicotinoid pesticides, which are neurotoxins, affect the bees' ability to navigate and remember, while fungicides can contribute to nosema (a disease that causes dysentery in bees). South African agriculture uses both, and while there is some suggestion that African honeybees have a stronger immune system than their European relatives, they are still susceptible to pesticides. But pesticide use is not the only problem facing bees in that region. Water is also an issue.

Water shortages not only complicate the lives of elephants and people but also create another environmental challenge that affects bees. Bees need plants, and plants need water to survive. This may seem obvious to most of us, but the ways in which this dependence plays out can be quite complicated. Some plants drink more water than others. For this reason, certain species—eucalyptus trees, for example—suffer extreme stress during drought. Water shortages have moved environmentalists like Guy Preston of the organization Working for Water to propose the removal of vast tracts of the invasive eucalyptus tree in South Africa, as they are large consumers of water. Beekeepers and bee researchers, however, worry about the effects of this proposal on bees. In a short documentary called *Bees*, Mike Allsopp of the Agricultural Research Council observes, "It is not an exaggeration to say that commercial beekeeping in South Africa began with eucalyptus plantations and, to a large extent, it is still dependent on them. Something on the order of 75 percent of all honey produced in South Africa is gum [eucalyptus] honey." But honey is not the only issue. In South Africa, the pollination of lychees, avocados, macadamia and other nuts, squash, pumpkins, celery, lettuce, carrots, cucumbers, berries, apples, pears, plums, and kiwi fruit all depend on honeybees. But a huge portion of the honeybees' yearly diet is the eucalyptus, the trees that are causing drought conditions, and without water, what will grow?

There are obviously no simple answers to this complex problem. Despite the many advances of technology, the search for a sustainable balance that allows human-nonhuman coexistence often feels like an elusive quest. Certain interventions, like the bee fences, seem to be one step toward achieving that balance. Interventions like pesticide use, which is another example of a human attempt to "control nature," seem only to further upset ecosystem health. Are human beings essentially or solely destructive resource consumers? Are there ways for humans to be a part of an ecosystem that allows for the survival of nonhuman others? And what about the unevenness within our human communities? For example, how and where are the San now?

In *Kalahari Communities: Bushmen and the Politics of the Environment in Southern Africa*, Robert K. Hitchcock paints a very dismal portrait of the quality of life for the San in recent years. Like many peoples of Africa, the San population has been affected by urbanization, land use and ethnic conflicts, and growing poverty and hunger. Hunter-gatherer people have been moved out of their homelands by development, making survival even harder than before. The impetus to protect land for charismatic fauna creates conflict for the indigenous groups. "One of the most significant problems facing indigenous minorities in Africa, according to a number of groups (e.g., Hadza, Pygmies, Bushmen)," writes Hitchcock, "is that they have been forced out of areas where they lived as a result of the establishment of national parks and game reserves."

Saving wilderness areas for species like elephants strikes most Westerners as a positive practice. But it also means that local people are displaced so that privileged Americans and Europeans can view these elephants from their Land Rovers. In looking specifically at the situation of the San, Hitchcock notes that many live in refugee camps or "freehold farms where they are visited by tourists, as in the case of Kagga Kamma," a private nature reserve. Poverty and unemployment rates are high. Where do bees fit into this picture?

One possible strategy for addressing unemployment and poverty has been designed by the Pretoria-based Agricultural Research Council in cooperation with the South African Department of Science and Technology, Department of Social Development, and Department of Agriculture. The project is called Beekeeping for Poverty Relief.

Beekeeping may be one of the few forms of agriculture with an overwhelmingly positive impact on the natural environment. The Agricultural Research Council sees it as a valuable conservation tool, allowing people to derive economic benefit from indigenous forests and other floral resources in a nondestructive way while ensuring local participation in conservation efforts. It also makes a significant contribution to other forms of agriculture by increasing pollination for many economically important plants.

The Agricultural Research Council aims to alleviate poverty while also encouraging sustainability. The long-term goal is to market not only bee products but beekeepers themselves. Ecotourists could come and see the process of beekeeping firsthand. The ARC website reports that "great interest is already being shown by at least one tourist operator, Sun and Sandals." One tourism website, purportedly dedicated to baby boomers, offers a beekeeping safari ranging in price from $5,000 to $6,000 per person.

While the efforts to alleviate poverty in such a sustainable and wholesome manner can be perceived as positive developments, it is hard to imagine that a people whose ancestors had such a rich body of knowledge, who could communicate with birds, who made sacred instruments like the bullroarer to bring the bees down, who knew how to read land and sky, are now going to be trained to perform a more domestic kind of beekeeping—and for tourists. And yet, this seems to be an overwhelmingly positive solution. For every decision or intervention, there are repercussions. Perhaps thinking mosaically will allow us to begin to resolve problems like these with the least destructive impact.

In a 1980 essay called "Ideas of Nature," the Welsh writer and social critic Raymond Williams articulated the desire for a shift away from the human-centered universe that had posited nature as something separate from and in conflict with the human being. Williams wanted to complicate this binary view of the world, and more specifically to challenge the separation of human and animal. He explored the impact that humans have had on "nature" and how we are inextricably tied to the earth.

> In this actual world, there is not then much point in counterposing or restating the great abstractions of Man and Nature. We have mixed our labour with the earth, our forces with its forces too deeply to be able to draw back and separate either out. . . .
>
> The process, that is to say, has to be seen as a whole, but not in abstract or singular ways. We have to look at all our products and activities, good and bad, and to see the relationships between them which are our own real relationships.

The activities and relationships to which Williams points have produced the earth we now inhabit. Elizabeth Kolbert's recent Pulitzer Prize–winning book *The Sixth Extinction* makes clear that this human impact is destroying species all over the globe. Illuminating the products and activities of our interspecies, "intermaterial" relationships is crucial for better understanding our impacts. We must actively examine the most destructive tendencies of human systems and be thoughtful about our interventions and changes. How to make the most widely ethical choices is the challenge we face. As we have seen in this chapter, different groups want different things and have different priorities. Some want to preserve wilderness and wild creatures; some, to ensure that human communities have enough money, food, and shelter; others, to save water; still others, to protect pollinators. The best solutions are those that take an interdisciplinary, interspecies approach. Bees are arguably "unnatural" in many ways (as they have been manipulated by years of breeding), but they are nonetheless essential participants in the world that we have built and now must navigate.

As I saw in China and in Africa, within this relationship between the bee and the human there exists a constellation of other relationships. Navigating changes and challenges involves being able to visualize a mosaic of actors. In looking at the situation of bees in southern Africa, I discovered new and surprising connections among flowers, fruit, water, artists, farmers, elephants, honey hunters, and policymakers.

In trying to understand how best to help the bees, I was realizing that there are no simple answers. Globalization invites universalizing solutions to our problems, but as Anna Tsing has pointed out, this universalizing pretends that the pieces that make up those universals are the same everywhere. There is no sameness, no evenness. Instead, there are countless shards. In this mosaic, I try to imagine all of the various and singular shards working together to create a new image, one I'm working to understand. Shards represented by beekeepers and biologists serve as individual pieces that do not lose their uniqueness in their coming together. Instead, in the course of gathering those pieces together, the relationships between them are brought out—those that jar the eye or the mind and those that seem harmonious. But out of this vast pile of shards, by imagining these mosaics, perhaps we can evaluate our impacts and relationships with more richness and even inspire innovative, interdisciplinary answers. I wondered what relationships and strategies I might now discover in my homeland. How were scientists, policymakers, farmers, and beekeepers in the United States reimagining their interconnectedness with bees?

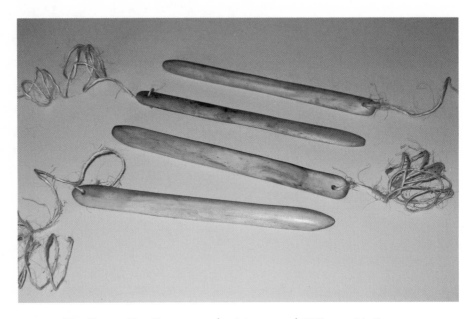

Kim Gurney, *Honey Hunters,* 2009 (work in progress). Yellowwood bullroarers carved from defunct beehives, shellac, sisal twine.

Making Sense of Bees
The Art of Kim Gurney

Kim Gurney is an artist, writer, researcher, and journalist living in Johannesburg, South Africa. Her multimedia installation and sculptural work has been shown in many galleries in Johannesburg and Cape Town. Two of her solo exhibitions, *Ecotopian States* and *Frugi Bonae,* explored the complex issues surrounding bee decline. Both of the shows utilized material used in beekeeping and employed a sensual understanding of the honeybee in order to encourage a sense of intimacy between the bee and the human viewer. In the exhibit *Ecotopian States*, for example, visitors to the University of Johannesburg Art Gallery were invited to listen to the haunting sounds of the bullroarer while standing in the midst of a group of bullroarers hanging from the ceiling, sculpted by the artist from defunct beehives. Gurney made a recording of the sound of this mystical instrument that had been used by the San as part of their honey-hunting practices.

Making an instrument to communicate with bees is a brilliant strategy. Through headphones, listeners were bathed in what could be described as a sacred aural experience. The sound bees make has captured me from the very beginning and was an important part of my early connection to them. The sound reminded me of the day I purchased my very first package of bees.

The roads I traveled to pick up my first package of bees were narrow, partly washed-out gravel paths through recently flooded fields in the upper Midwest of the United States, a long, long way from South Africa. The rural landscape of Wisconsin was still barren after a very cold winter. The sky was the color of hammered steel, but my mind was filled with images of flowers and bees and honey. The bee equipment-supply company was located in a large warehouse. The woman at the cash register sent me out back to get my bees. Behind the building was a semi-trailer truck full of *Apis mellifera*. A burly man in a plaid flannel shirt and well-worn, golden Carhartt pants was standing inside the trailer. I told him I had ordered a package of bees, and he walked back into the darkness and returned with a box a bit smaller than a shoebox. The "box" looked like a miniature old-fashioned box kite, with thin wooden boards articulating edges of the walls of flexible wire screen. The squares of wood that formed the top and bottom were the only parts of the container that seemed particularly strong. And inside were ten thousand bees. "This is my first package of bees," I said to him nervously. "Anything I should know?"

He laughed. "Don't go over any big bumps on the way home!" And he turned and walked back into the trailer.

I stared through the mesh at the insects. They were clustered together in a large clump around what I knew had to be the smaller package that held the queen. I carried them carefully back to my Jeep. I was imagining what might happen if the box did break open. Would I have time to get out of the vehicle without being stung to death? I had read somewhere that it would take ten stings per pound of body weight to take a person out. I wasn't great at math, but I knew there were enough bees in there for that. What had I been thinking? This was pretty legitimate danger. I put the box on the floor of the passenger's side and walked over to the driver's side, got in, and closed the door. It was in that small, quiet space that I first heard them humming. The sound was low and throaty, calm and gently undulating, a sound not unlike a very quiet version of Zen monks chanting. My fear dissolved. This song, this meditative sound the bees were making, communicated to me very quickly that these creatures had no interest in harming me. I drove for a long while listening to their humming. And finally, I began to sing, hoping they would understand by the tone of my voice that I also meant them no harm.

Of course, I eventually learned that as a human who ate food and lived the average Western lifestyle of consumption, I was certainly complicit in their demise, but what I also learned was that they would not hurt me unless I did something egregious to them—like pinching them with my rushing fingers when I was opening the hive too quickly, or approaching during robbing season, when they were being attacked by other bees trying to steal their honey. The stings seemed justifiable. Our abuse of them far exceeded the stings I have endured. Would people coming to an exhibit of Kim Gurney's begin to understand our role in the disappearance of bees?

Kim Gurney's use of the sound of the bullroarer created an embodied experience for the viewer, which could produce a new connection with the bees that went far beyond the visual or intellectual. For many people, I'm sure, this sound was not an experience of the sacred, and perhaps it even produced fear in some, but for others the sound created a perspective shift that was unique and offered a new avenue for knowing. Gurney herself writes,

> My core idea was to create something "non-Western"—so it is looped, without a resolved beginning, middle and end in the classical formal musical structure. Instead, it has a kind of throbbing, pulsating breath-like timbre to encourage a low-grade anxiety in the listener through a hyperventilating rhythm, created by the bullroarer swinging action. . . . It is very deliberate that the piece can only be listened to by a single person at a

Kim Gurney, *Honey Hunters,* 2009 (installation view). Yellowwood carved from
defunct beehives, shellac, sisal twine, Perspex, and audio, approx. 3,000 × 1,000 mm,
4'4" audio (looped). Exhibition view from University of Johannesburg Art Gallery.

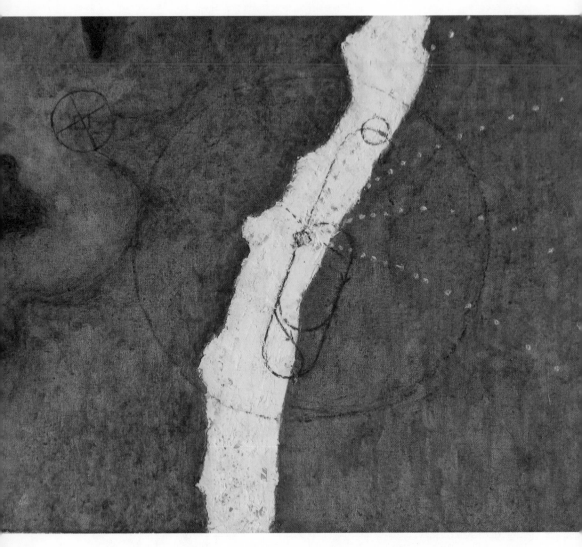

Kim Gurney, *Bringing It Down I,* 2010. Oil, beeswax, and iridescent medium on canvas, 45 × 55 × 4.5 cm. Private collection, Cape Town.

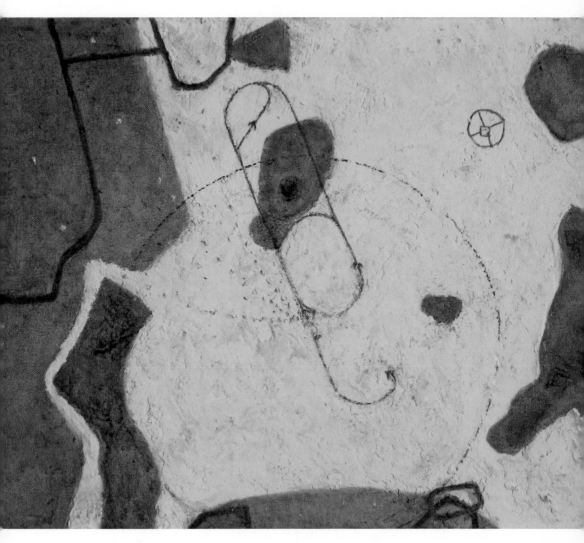

Kim Gurney, *Bringing It Down II*, 2010. Oil, beeswax, and iridescent medium on canvas, 45 × 55 × 4.5 cm. Private collection, Cape Town.

Kim Gurney, *Still Life: Pears*, 2009. Aluminum honeycomb panel, nuts and bolts, Fabriano, bee smoker stencils, 105 × 31 × 4 cm. Private collection, Johannesburg.

time—it's a solitary experience, and in a fairly constricted "booth" encircled by the stationary forms of the actual "bullroarers" that performed the sound. The effect is hopefully quite bodily.

Other pieces in this show played with the idea of landing guides. The series *Bringing It Down* is a meditation on the ultraviolet landing guides bees use to detect nectar on flower petals and the landing schematics humans use to land airplanes. This series illuminates yet another form of multispecies communication—that of honeybees

and plants. Flowers have evolved to create these vibrant landing strips for the bees. But many of the flowers bees depend on are disappearing because of development and drought.

In another show, called *Frugi Bonae*, Gurney expanded her work on the bee. In one piece, *Still Life: Pears*, she explores the idea of "perfect fruit," which is demanded by "first-world shoppers." Fruit with any imperfection or variation is not considered "perfect," and many insecticides are used to achieve a perfectly colored, perfectly shaped, unblemished appearance. As we

have seen, this rampant use of pesticides has profoundly negative effects on many other life forms, including bees. Gurney used a bee smoker to create stencils of perfect fruit that serve as a backdrop for what appears to be a ribbon of burned honeycomb. The mangled comb floats precariously inside the frame, suggesting the devastation that honeybee hives are currently enduring. The burning of comb and hive has been employed by many beekeepers who are trying to fight American foulbrood disease, the most widespread and destructive disease bees face. These highly contagious bacteria can be completely destroyed only by fire. One theory about the spread of bacterial and viral diseases is that it is caused by the high traffic of bees, not only from one industrial crop to another but from country to country. When bees (or queens) are scarce in one country, bees will be shipped over international borders. The bees are thus combatting both a host of diseases and toxic chemicals on the flowers they need for sustenance.

In *Labour of Love II*, Gurney examines the widespread practice in commercial apiculture of trucking hives of bees long

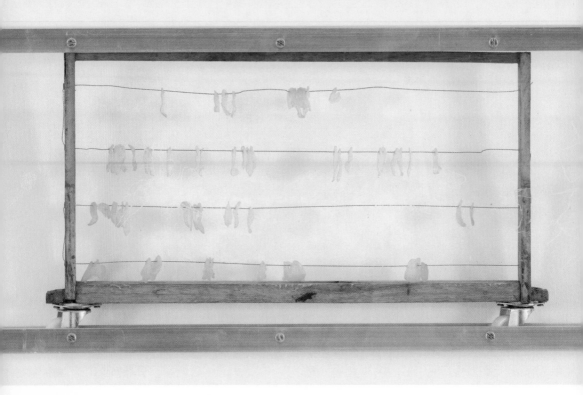

Kim Gurney, *Labour of Love II* (detail), 2009. Beehive frames, wax petals coated with shellac, aluminum rails, casters, nuts and bolts, engraved Perspex, 250 × 30 × 3.5 cm. Private collection, Johannesburg.

distances to pollinate vast orchards of agricultural crops. "Dismembered beehive frames (where honey is created and stored)," Gurney writes, "are presented on casters along an aluminium production line. The frames are re-contextualised as a kind of ecological abacus for artificially coloured wax petals that hang in delicate balance." Gurney's work invites us to examine the precariousness of our current relationship with bees, as we have reduced their bodies to function as tools for industrial agriculture.

The wide variety in Kim Gurney's work is united in its attempt to create new methods of expressing the situation of bees that are not purely representational. She asks us for a more embodied, intimate connection. How was the work of scientists affected by knowing bees intimately? For Sainath Suryanarayanan, it meant shifting his career.

Killing Bees to Save Them

THE HEARTBREAK OF EXPERIMENTAL DESIGN

chapter four

It was a warm day in early spring when I had my first long conversation with the entomologist and science studies scholar Sainath Suryanarayanan. We met over a couple of hives I had recently inherited. One was thriving. Piles of dead bees filled the other. Parts of the comb were covered with mold and oozing something that looked like molasses. Having recently attended a class for hobby beekeepers with Marla Spivak, an entomologist at the University of Minnesota, I was aware of the many different diseases to which bees are susceptible. American foulbrood, one of the meanest, concerned me most. Beekeepers recommended burning all of your equipment if you discovered it in your hives. Some of these bees were alive, but obviously in low spirits, and I didn't want to destroy them unnecessarily. I called Sainath because I thought he could help me with the diagnosis.

Beekeeping, these days, is obviously riddled with risks. New viruses, habitat loss, pesticides, and mites all contribute to creating a deadly labyrinth through which nearly every bee must travel. Diagnosing bee problems is not simple, but some answers seem to be emerging. A beekeeper facing a failing hive now has to consider not only the health of the hive itself, but also the health of the landscape around the hive. Dead bees lead beekeepers down a path of many questions. And some beekeepers have lost so many hives, they feel like giving up.

When we met at my troubled hives, Sainath brought his own hive tool and veil. He had already been down a path of many questions about bee deaths, one that started in his youth with his fascination for observing insects. At age fourteen, he began keeping an "Amateur Entomologist's Record," where he made taxonomic notes on such things as wing textures, body shapes, color patterns, and behaviors. But the young scientist's approach occasionally slipped to include his exuberance; he described one moment as "a stupefying experience!" Sainath studied biology and chemistry in college, then focused on the behavioral ecology of paper wasps during his doctoral studies, and eventually moved to Minnesota, where he helped Spivak investigate the role of pesticides in CCD.

Sainath had spent several years doing lab and field experiments with wasps and bees, but he ultimately wanted to shift from traditional practices in entomology to research that included human/insect relationships. It was Sainath who made me wonder about the role of emotion in science—both in the scientists themselves and in the subjects of their experiments. I had always thought of emotion as something excised from science, but this was impossible for some scientists. What was the role of empathy in experimentation? How do we, with our human limitations, understand something as radically different from us as the honeybee? Did bees have feelings, too? If so, what did that mean for the scientist? For science?

Sainath's gentle demeanor and velvet voice seemed to be balm to the bees, and to me as well. We cracked open the dead hive and were immediately engulfed in the smell of death. Even before we could see the bee bodies, it was clear that there was something terribly wrong with this hive. Beehives usually have an intoxicating smell. The reason for this may be obvious: that honey and beeswax and propolis, the products of creatures trafficking in nectar and pollen, would probably smell as good as the flowers they frequent. But bees are also great housekeepers.

When bees first emerge from their wax cells, they begin their lives doing the work of hive maintenance. They patch holes with propolis (which they harvest from certain trees and flowers), repair damaged comb with wax, and drag tiny unwanted objects—dirt, twigs, dead bees, or other insects—out of the hive through the main entrance. This is why there will often be a few dead bees at the entrance of even a very healthy hive. When an object is too large or impossible for a bee to remove, she will encase the object in propolis to keep it from polluting their sacred space. But the piles of decaying bee carcasses in this hive clearly indicated something terribly out of balance. Sainath and I began clearing away the debris of the devastated city.

In the Hindu tradition in India, where Sainath grew up, the bee is symbolic of both the divine and the sensual. According to one bee blogger,

> The god Indra was the namesake of ancient India and the deity who separated heaven and earth, and is said to have received honey as his first food. Similarly, the Indian Bee goddess Bhramari Devi derives her name from the word *Bramari*, meaning "Bees" in Hindi. It is said that Bhramari Devi resides inside the heart chakra [the seat of compassion] and emits

the buzzing sound of Bees, called "Bhramaran." Likewise, the sound of a Bee humming was emulated in Vedic chants, and the humming of Bees represented the essential sound of the universe all across India.

When Sainath was a little boy, he captured and made careful observations of insects, not because he considered the insects sacred but because their complexity fascinated him.

Sainath and I carefully took the hive apart, frame by frame, and decided that it didn't actually look like foulbrood after all. We did find more mold, which we thought might have been caused by water that had leaked into the hive. Could a leaky feeder have been the culprit? In a climate like ours, in the midwestern part of the United States, beekeepers often feed their bees in early spring, before blossoms open, with something called "spring syrup," a concoction made of sugar and water. One feeding method involves puncturing the lid of a Ball jar and then placing the jar inside the hive upside down. If the jar is not resting evenly or the holes are too large, the syrup can seep out more quickly than the bees can drink it up, and the result is a wet hive. Could the previous owners have had a leaky feeder? The damage was bad, though, and very little of this hive could be salvaged.

A year or so after examining my damaged hive, Sainath and I met to talk about the suffering of the bees. He began by telling me about his work with paper wasps. His research proved that socially dominant female wasps participate in drumming behavior to create a nest vibration that alters the development of baby wasps. He held an imaginary pipette between his thumb and middle finger. To track their development, he said, "we had to puncture the bodies of the live babies and suck their blood like this." He held the imaginary pipette to his mouth, puckered his lips, and sucked. His finger capped the end and carried the imaginary blood to a vial. "Over and over we did this, and then the dreams began." His dark eyes darted to the floor, his brow furrowed. "Yes, they suffer."

The seventeenth-century French philosopher René Descartes suggested that animals are like machines and lack the capacity for thought. For many years, those who attributed any kind of emotion or reasoning power to an animal were widely ridiculed. Lorraine Daston and Gregg Mitman, both award-winning historians of science, summarize this fear of anthropomorphism in the introduction to their book *Thinking with Animals* (2006): "'Anthropomorphism' is the word used to describe the belief that animals are essentially like humans, and it is usually applied as a term of reproach, both intellectual and moral." Anthropomorphizing has been seen as bad, sloppy thinking, a view that Daston and Mitman seek to overturn.

Sai leaned forward into the small circle of our quiet voices, the small circle of intimacy created by the fact that we both knew how many people would find our view of insect suffering blasphemous. Our eyes locked. "I didn't prove it *scientifically*," he said, "but I could feel it, you know. Of course they could suffer. Their long antennae caressing each other or on my hands—their feet and tongues touching. Yes, they do. They suffer."

I felt a tightness in my chest as I imagined my own honeybees, their tiny faces, the soft yellow hair on their abdomens, the tentative way they explored my hands. The way they touched each other with so much tenderness. I could not imagine piercing the belly of a live bee with a scalpel, watching it struggle and flail. I feel a kinship with Virginia Woolf, who describes the struggle of a moth in her short meditation "The Death of the Moth." "Watching him, it seemed as if a fibre, very thin but pure, of the enormous energy of the world had been thrust into his frail and diminutive body. As often as he crossed the pane, I could fancy that a thread of vital light became visible. He was little or nothing but life." Woolf turns away, but then looks back and realizes that the moth is struggling against death:

> . . . when he tried to fly across it he failed. Being intent on other matters I watched these futile attempts for a time without thinking, unconsciously waiting for him to resume his flight, as one waits for a machine, that has stopped momentarily, to start again without considering the reason of its failure. After perhaps a seventh attempt he slipped from the wooden ledge and fell, fluttering his wings, on to his back on the windowsill. The helplessness of his attitude roused me. It flashed upon me that he was in difficulties; he could no longer raise himself; his legs struggled vainly. But, as I stretched out a pencil, meaning to help him to right himself, it came over me that the failure and awkwardness were the approach of death.

Of course, insects are not machines, though I know that many people consider them tools, not sentient beings.

Anyone who has observed bees has surely intuited their enormous sensitivity to touch, but Mariana Gil and Rodrigo J. De Marco, from the Institute of Cell Biology and Neurobiology in Berlin, wanted to prove something more—namely, that bees communicate cartographic information through touch. As many of us know, the waggle dance of the honeybee, first decoded by Karl von Frisch, is the language bees use to communicate the distance to a nectar source or the location for a potential new hive. Bees dance a map by relating the azimuth of the sun to gravity inside the hive . . . in the dark. The bees register this information through their antennae and translate it in their brains. "Using high-speed video techniques," say Gil and De Marco,

> we found that the higher the number of the dancer's wagging movements, the higher the number of the followers' antennal deflections. We also documented that most followers faced the dancers laterally and experienced a fairly regular pattern of tactile stimuli; a much smaller proportion of followers faced the dancers from behind and became the subject of a different, although still regular, pattern of tactile stimuli. From these observations, we conclude that tactile mechanosensory input from the antennae, presumably processed by neurons of the antennal joint hair sensilla and the neck hair plates, enables bees to estimate both the direction relative to gravity and the length of the waggle phase.

This experiment proves how much is communicated through touch alone. These insects are immensely sensitive.

But what is that experience of touch really like? Does it translate to something like human pain or emotion? This has been a very controversial issue for centuries. Marc Bekoff, an animal ethologist from Colorado, has spent his career thinking about human relationships with animals, and about whether animals have emotions. "It's bad biology to argue against the existence of animal emotions," he observes. "Scientific research in evolutionary biology, cognitive ethology, and social neuroscience supports the view that numerous and diverse animals have rich and deep emotional lives." From Bekoff's perspective, even if we cannot map these feelings directly onto our own experience, they still matter.

Many animals experience pain, anxiety, and suffering, physically and psychologically, when they are held in captivity or subjected to starvation, social isolation, physical restraint, or painful conditions from which they cannot escape. Even if it is not the same experience of pain, anxiety, or suffering that humans undergo—or even that other animals, including members of the same species, feel—an individual's pain, suffering, and anxiety matter.

Bekoff has studied elephants, who exhibit a form of posttraumatic stress disorder upon seeing other elephants violently killed, and has observed the compassion of whales and primates, ritual mourning among magpies, and interspecies friendships between such strange mates as a snake and a hamster. He wonders whether even insects might have emotions.

Over the years, at my own hives, I have sensed moods, certainly. The bees in the hive that Sai and I opened together seemed to lack vigor and—dare I say—hope. We decided to destroy the blackened frames, just in case it was foul-brood, by fire. But there were survivors. I put those bees on clean frames and their energy lifted. Within a week or so, I realized that something was still amiss, because their energy seemed frenetic. When I checked the frames, I found no eggs and no baby queen cells, which meant that there was no queen present or soon to be born. A queen is essential; she is the mother of all the new workers in the hive. She lays, on average, one to two thousand eggs a day. Without a queen, the hive would die. I quickly called some beekeepers to see if anyone had any queens left, and, luckily, a friend of a friend was able to send one to Wisconsin from Florida. She and her small entourage of attendant workers arrived via UPS in a small makeshift box with sides made out of screen. The queen herself had her own compartment, a wooden box about the length of my thumb that had a screen on one side and a small hole filled with a piece of hard candy at the end, an edible barrier. Her attendants were clustered around her in a clump about the size of my hands clasped together.

When you "re-queen" a hive, there is always a chance that the bees will kill the new queen if they do not accept her. The way to avoid this is to place the new queen in her tiny compartment inside the hive for at least a day or so, so that the bees will grow accustomed to her pheromones before you release her. I placed the tiny container inside the hive, whispered a few words of introduction to the struggling community, closed up the hive, and crossed my fingers. A day later, I pried off the hive cover, carefully pulled back the screen wall of the queen's chamber, and released her into the hive. Immediately, seven or eight workers surrounded her like petals of a sunflower and began stroking and grooming her, and I felt a wave of relief. Within a few days, I perceived the mood of the hive to be calm, happy, and busy, at last.

But did bees really have emotions? Some new science suggests that in fact they do. An article in *Current Biology* called "Agitated Honeybees Exhibit Pessi-mistic Cognitive Biases" discusses the discovery of "proof of emotion" among honeybees. Researchers at Newcastle University have been interested in mapping emotions in nonhuman species. They have worked for a long time with vertebrates but decided to see if they could detect emotion in honeybees. The experiment they designed involved training bees to associate certain scents with either rewards of sucrose or "punishments" of bitter quinine. The bees are captured and slipped into tiny tubelike harnesses that keep their wings from moving while their heads stick out one end. Scientists watch the faces of the bees very carefully. A bee who smells the scent associated with a reward sticks out her tongue. A bee who smells the scent associated with bitterness retracts her mouthparts. This response is so reliable that bees trained in this manner can be used to detect drugs or explosives. The scientists at Newcastle wanted to know whether stress would affect the bees' decisions, so they vigorously shook one group of bees to simulate an invasion of the hive. The bees, it turns out, were not just physically but psychologically shaken. Not only did they produce lower levels of serotonin and dopamine; they also became pessimistic in their responses and assumed they would be given more punishment. This pessimism is seen as an emotional response to stress. This thrilled the researchers, who concluded that "the bees' response to a negatively valenced event has more in common with that of vertebrates than previously thought. This finding reinforces the use of cognitive bias as a measure of negative emotional states across species and suggests that honeybees could be regarded as exhibiting emotions."

None of this proves conclusively, however, that honeybees experience pain, emotional or physical, that is similar to human pain. There is no way for us to know that, and what would it matter if we did? Historically, the general consensus and acknowledgment that certain animals (particularly certain mammals) experience pain have given rise to animal rights movements and anticruelty laws. But most of those laws are often enforced unevenly in the lab. In a 2008 article, the social scientist Kristin Asdal focused on a historical case in Norway that exemplifies how societies have negotiated boundaries with respect to animal treatment in labs. In spite of the animal rights movement and a series of laws designed to protect (certain) animals from abuse in the public sphere, scientists were given permission to do whatever they liked to lab animals. "Whoever . . . should be guilty of gross or malignant mistreatment of animals," states the Norwegian law, "or whoever aids or abets such an act, will be punished by fine or imprisonment up to 6 months. This decision does not hinder the King, or someone to whom the King has bestowed authority, from allowing appointed persons in designated places to conduct painful experiments on animals for scientific purposes." In the lab, the animal is a tool in the service of science, a means to an end. It is important to note, however, that scientists working with vertebrates today must follow certain specific protocols that govern how to handle or kill an animal in the lab. Failure to follow these protocols can result in a loss of funding or the termination of a career. Still, scientists are not subject to the same rules and laws that apply to the public. And laboratory experiments on invertebrates lack such protocols. (Killing animals for meat is another issue, governed by its own rules, but this is an issue for another discussion.)

Strangely similar to the use of the bee as a sacred symbol, the use of bees in the lab ignores the actual life of the bee, separating its ontology from its function as a tool or symbol. Society has essentially condoned the infliction of pain in the laboratory, as long as the purpose is integral to the betterment of human society. The lab is a specialized and sacred place where sacrifice is an acceptable part of the system.

Most of us cheered in the movie *E.T.* when Elliott, inspired by his relationship with the adorable extraterrestrial, liberates the frogs in his school's biology lab. Animal testing is not something we like to look at or think about, but most agree that it is necessary for advances in medicine and biological knowledge. But what about the scientists who work with these creatures? How do they shut down their own emotions and their sense of compassion in order to continue working?

Lorraine Daston and Gregg Mitman, in their introduction to *Thinking with Animals*, describe a kind of bond building between scientists and animals: "ethologists who have devoted their lives to the study (and often the preservation) of elephants, gorillas and other at-risk animal species develop deep identifications with their chosen subjects. Among scientists who investigate animal behavior, such feelings are not uncommon, and even those who disapprove of anthropomorphism in ethology in principle admit that in practice the arduous life of the observer in the wild would hardly be tolerable without some such emotional bond."

So what does a scientist do with this emotional bond when hurting or sacrificing animals, even insects, in the lab? The suffering that Sainath inflicted on his paper wasps caused him immense suffering. Over time, his visceral nightmares grew worse and worse. He often woke up gagging from dreams about drawing blood from the brood. His work with Spivak's team of entomologists at the University of Minnesota bee lab proved arduous, too. Charged with testing pesticides, especially imidacloprid (one of the neonicotinoids), he spent nine hours a day checking the health of bees, looking for changes in behavior and brood-development patterns; and nine hours a day he slowly poisoned the bees as well. The U.S. Environmental Protection Agency wanted proof of the chronic toxicity of imidacloprid in beehives. "That meant we had to kill them," said Sainath. "They became sick and grew sicker and sicker until they would finally die." The sugar syrup and pollen substitutes that Sainath fed the bees were laced with progressively higher doses of the pesticide. "Marla [Spivak] and I were uneasy about poisoning the beehives," he said. Not only was it wrong to be killing healthy bees, but this was not the way they would encounter the pesticide in their natural environment.

The testing was based on a toxicity standard called an LD50 (lethal dose 50 percent), which was measured by the amount of a chemical the bees could ingest before 50 percent of them died. What became increasingly clear to Sainath was that focusing on how much of one chemical it takes to kill a bee was not good enough. The research failed to account for other, interacting factors—synthetic chemicals, pathogens, and nutrition, to name just a few. It did not test far smaller pesticide doses that could, over time, damage the bee brain and its ability to sense and navigate the world.

Above all, the research did not examine the many human-honeybee entanglements that occur in the real world—a vast oversight given that we ourselves, the beekeepers and growers and all of our systems, are profoundly implicated in the bees' decline. The chemical regimes of industrial agriculture are deeply entrenched and highly complex, and the forced movement of millions of hives for the purpose of pollinating monocrops, leading to the loss of wildflower habitats, represents an associated loss we must account for as well.

But if it was clear that these toxic chemicals were making bees sick and making them suffer, why didn't the EPA just eliminate them? To learn more about the EPA's regulation of pesticides, I contacted the entomologist Kaitlin Stack-Whitney, who worked for the EPA during the years that Sainath was conducting his pesticide experiments in Minnesota; she was eager to talk with me. I met her in a bright café on a cold day in November. Unlike Sai, Kaitlin had not loved insects as a child, but she admitted that the only pets in her suburban Rhode Island home were the ants in her ant farm. Raised on stories of her Irish and Italian ancestors, who had come to the United States because of famine in their own countries, Kaitlin became interested in international agriculture and rural development, which she eventually studied at Cornell. Working in a campus dining hall to make extra money was not entirely satisfying, so when an opportunity to work as a technical assistant in an agroecology lab came up, she took it. It was there that she first really connected with insects.

Kaitlin's brown eyes brightened when she talked about aphids. Her job at the lab was to keep four colonies of the tiny creatures alive. Aphids are sap-sucking insects whose average length is one to ten millimeters. "They were so cute," she said. "They have little green, round antennae, tiny little eyes, and cornicles on their butts that emit chemical cues to each other for communication." She was in charge of millions and millions of aphids who were kept in growth chambers with a strict temperature and light regime. The aphids were disease vectors among grain plants, but the viruses the grains carried did not affect the aphids themselves. With a fine-tipped paintbrush, Kaitlin would place ten aphids on one leaf per plant. The aphids had to be placed facing skyward or they would crawl off the plant. This, as you might imagine, was a painstaking process. Aphids are capable of parthenogenetic reproduction, which means that the females can clone themselves and create telescoping generations. This allows them to double their population in a few days. It also makes aphids a huge pest for farmers. "They are adorable," she said, "but there are thousands of kinds, and they carry thousands of viruses."

Inspired by the aphid work, Kaitlin began thinking about food webs and developed a thesis that required her to search the cornfields of six counties in upstate New York for "natural enemies" of the corn farmer's most tenacious pests: corn borers and corn rootworms. Could you use natural enemies rather than pesticides to suppress pests? After working for the USAID Farmer-to-Farmer program, which helps grower collectives around the world, she landed a job at the EPA working on pesticides, and her question about nontoxic pest control had to be set aside until a later date.

"The EPA is not an activist agency," Kaitlin explained. "They enact authority for pesticides." Half of the staff are scientists, the other half are economists, and together they work with a cost-benefit model. This is much different from the European model, which uses the precautionary principle, which allowed the EU to try a temporary ban on neonicotinoids. Every new pesticide in the United States has to be registered with the EPA; the burden of proof is on the company that wants the registration. Testing is done by third-party labs, and the company pays the regulators. Taxpayers, Kaitlin asserted, should not be paying for this testing. This is the only part of the EPA that is a licensing agency, and so it deals directly with industry. "We need thorough information about the costs and the benefits from the companies, so there is obviously a tension." The EPA wants to know about human health considerations: the effects of pesticide residues on food, in breast milk, and in subsistence diets, and the effects of dermal contact for farmworkers, applicators, and consumers. It also wants to know about "downstream effects"—where pesticides go after they leave the fields—so it looks at runoff, at the effects in drinking water and on frogs and fish, and at how the chemical breaks down in soil. Then EPA regulators weigh the costs against the benefits. Is the pesticide cheaper, safer, more effective than alternatives? How many frogs will die? Will it create greater yields for farmers? I imagine that coming to a conclusion must be a very long and difficult process.

It is not easy to get an approval, Kaitlin said, but getting rid of a pesticide once it has been accepted is much harder. An approved chemical is legally bound to be retested every fifteen years, but the burden of legal removal is now on the government. In order to revoke a pesticide, the government must demonstrate that its use in every single U.S. state cannot be mitigated or restricted to create safer usage. The EPA suggests shifting doses, the timing of application, the pounds applied per acre, and the number of applications per season. This becomes very complicated when you consider that each state has its own regulations. California, for example, has "townships" in which certain chemicals that create lifetime cancer risks are regulated. This means, Kaitlin explained, that you can use only a certain amount of that chemical over the course of an average lifetime or, in other words, that you can periodically use quite a bit and then cut down. "That is very spatially sophisticated," she said. Reducing usage would make the chemical "safer," and unless the chemical proves to have truly terrible effects even in "safe" amounts, that chemical will remain legal. To cancel registration or to eliminate a chemical from use in the United States takes years.

Kaitlin grew quiet, and then began a string of sentences that felt in their tone like an apology. "The good thing is the transparency of the process. Each pesticide has a docket of its own. Scientists need to sign off on all the documents. There is a public comment period. It's not easy to get rid of things once they're in, but it's hard to get them in there in the first place." She sighed. "The problem is that the process is not working fast enough for issues around neonicotinoids." The EPA abides by the rules it is required to follow. The agency is not required to listen to beekeepers who think their bees are dying from pesticide poisoning. If you want to protect the bees and the environment, she said, you need to change the rules.

As Michel Foucault suggests in *Discipline and Punish*, embedded knowledge is always contaminated by power. Our government privileges scientific knowledge over the experience of beekeepers. The EPA is, by design, enmeshed with the agrochemical industry. Bee advocates face this imbalance of power as they confront Colony Collapse Disorder.

Sai Suryanarayanan and his colleague Daniel Kleinman are trying to shift the balance. Their work exposes the ignorance that perpetuates the use of toxic substances by large agrochemical corporations.

Neonicotinoids in small doses are destroying the bees' ability to navigate, to smell, and to remember. Because bees use the position of the sun for navigation, dance a map for the worker bees inside the hive, and identify the smell of specific flowers on the bodies of other bees, they need all of their senses to make the decisions that keep the hive humming. A healthy nervous system in each bee is crucial to overall hive health. Neonicotinoids make the bees incapable of functioning normally. Unfortunately, when organizations like the EPA decide to privilege certain methods of scientific inquiry, they condone the continued use of chemicals that are killing billions of bees.

Sainath and Daniel Kleinman aim to change the way we measure toxicity and how we determine the safety of a chemical. The trickiest part of the EPA system is providing conclusive evidence that a specific chemical (or group of chemicals) is harmful. One obstacle to proving this is that most experiments test the effects of only one chemical at a time. When bees fly out into a field, they might encounter a whole cocktail of pesticides. How would you test for that? What if some of the other chemicals are lowering the bees' immune systems? Fungicides, for example, increase the incidence of nosema, a kind of dysentery in bees. Sai believes that some of the most important voices in the discussion about new ways of understanding bee decline belong to beekeepers, people who spend their lives paying attention to bees. Sainath and Daniel argue that "dominant approaches to honey bee toxicology institutionalize particular kinds of ignorance about the involvement of agricultural insecticides in CCD. This ignorance, in turn, justifies a lack of regulatory action on the part of regulators at the US Environmental Protection Agency (EPA) and ultimately serves the interests of elite agrochemical companies."

Sainath and Daniel believe that beekeepers, who are very careful observers of bees and their habits, can offer a great deal of insight into the problem of CCD. While the beekeepers were noticing that bees were not doing well after their contact with neonicotinoid pesticides, the chemical companies claimed that those pesticides were safe. In an article in the *Guardian*, they write:

> Despite the conclusion of beekeepers across the globe, based on their field research, that neonicotinoid insecticides were likely contributing to increased bee mortality, some chemical company representatives, scientists and government regulators dismissed or disparaged their findings. Our view is that commercial beekeepers have real-time observational knowledge of the crisis facing honey bee pollinators and that we should take their research seriously. . . . Our point is not to say that commercial beekeepers always know best. Rather, it is to argue for more genuinely participatory research that brings beekeepers' knowledge and scientists' knowledge into a creative and egalitarian dialogue toward a fuller understanding of why honey bees are dying.

In September 2013, Suryanarayanan and Kleinman launched a "transdisciplinary inquiry" funded by the U.S. National Science Foundation involving the range of stakeholders on the ground—entomologists, ecologists, beekeepers, growers, government representatives, and humanitarians. In collaboration, they have designed experiments to save bees without tormenting them. In one approach, they've studied beehives in a variety of landscapes with varying amounts of pesticides, and then sampled pollen to see which communities are most effective, and why.

How might a beekeeper, who knows when bees visit particular flowers, help refine a study's design? How can the knowledge of an ecologist help interpret those findings? How could a grower, who knows the timing and methods of chemical applications, help analyze these results? These conversations will eventually—it is hoped—affect policy and future research.

The summer of 2014 was the first season of Sainath's alternative approach to bees, and he started by observing them in the landscapes in which they lived. As a research assistant to Sainath and his diverse team, I visited bees living at the edges of meadows full of wildflowers and hives nestled on the edges of corn and vegetable fields. We spent time talking to farmers and beekeepers about their habits, approaches, and concerns.

The life of honeybees cannot be divorced from the humans who work with them, count on them, love them. We opened hives and took note of their activities and their health. As we lifted the frames, Sainath whispered soothing words to the bees. As with his previous work, the new approach provokes complex emotions, but his nightmares are gone.

In an article that explores the "emotional ecology of field science" that occurs in multispecies relationships among scientists, horseshoe crabs, and red knots (a northern shorebird), science and technology scholar Kristoffer Whitney writes, "Empathy, wonder, and technological and methodological enthusiasms make science in the Delaware Bay happen, and this is visible in the prosaic experiences of wildlife researchers and managers there. These scientists move, however, from research sites on the bay into multiple public, decision-making bodies, and these movements require the translation and obfuscation of the values underpinning their work bayside into forms socially acceptable in the context of rational bureaucracy." Basically, Whitney suggests, we are not allowed to reveal the emotional valences that inform policymaking or the science it is based on. But science is informed already by emotion, by compassion, as it must be. Aren't decisions that are made without real connection to those they will affect apt to be poor ones? How disconnected can we afford to be when faced with the disappearance of an insect crucial to the pollination of one-third of the food we eat?

Sainath's uncommon compassion for insects, his unusual connection with them, motivated his innovations in experimental design. For Sarah Hatton, a Canadian painter, compassion for insects also motivated a professional shift. In this case, Hatton abandoned paint and began gathering bags of dead bees.

Sarah Hatton, *Circle 1*, 2013. Honeybees (*Apis mellifera*) and resin on panel.

Mourning the Dead
The Art of Sarah Hatton

Sarah Hatton received her BFA from Queen's University in Ontario and her MFA from the University of Calgary. When this artist and beekeeper lost all of her bees, she was devastated. For many years, Hatton had been best known for her brushy, expressionistic paintings depicting children playing and pondering in the water, waterbirds, and dogs in dripping blues and greens, and for her explorations of how light is broken by tangled branches. When all of her bees died suddenly, however, Sarah's process of mourning pushed her into a new artistic realm. What could she do with all of the dead bees to tell the story of their fragility?

Rather than pick up her brushes, she collected the dead bees and began arranging them on boards. Using the carcasses of the bees may seem morbid, but the practice of preserving dead animal bodies is not a recent phenomenon. Egyptians embalmed animals. Hides, bones, and feathers have been used by many groups of people for ritual and clothing throughout human history. The specific practice of taxidermy has been around for centuries. Taxidermy collections were particularly popular in

the Victorian era, when everything from hummingbird cabinets and game hunters' trophy halls adorned collectors' homes. Natural history museums are filled with stuffed animal bodies. In her essay "Teddy Bear Patriarchy," Donna Haraway walks the reader through the American Museum of Natural History in New York City, where curators have created dioramas featuring the work of taxidermists like the famous Carl Akeley. Inside the exhibits, Haraway narrates the meanings that this museum makes, arguing that the "natural" dioramas, the constructed scenes of unadulterated nature made possible by the heroic act of killing, inscribe lessons of dominance and traditional Western ideas of white masculinity.

Giovanni Aloi, the author of *Art and Animals* and the founder and editor-in-chief of *Antennae: The Journal of Nature in Visual Culture*, is interested in the use of taxidermy in contemporary art. Partly in response to Haraway's ideas, he became curious about the ways in which some taxidermists are now consciously grappling with a different social and cultural context. "I am trying to resist the production," he says in a recent interview, "of 'sweeping readings' of phenomena such as taxidermy or the presence of live animals in the gallery space in order to favour more fragmented and dissonant configuration of the contemporary scene. ... We are perhaps witnessing the emergence of a Post-postmodern animal or an Altermodern one in which the relationship between animal and artist is increasingly based on a personal relation rather than on the metanarratives of the past." Aloi's interest in taxidermy led him, in fact, to dedicate an issue of *Antennae* to a group of contemporary taxidermists, about whom he said, "the return of taxidermy is not a hollow revival but a highly intriguing phenomenon. Victorian taxidermy played a very different role within its socio-historical scenario from that played by the contemporary art revisionist movement." "Contemporary taxidermied art has its own agenda," he continues; "interestingly, this is one set almost entirely by women. They have embraced the ancient craft and are determined to subvert the 'masculine' idea of the taxidermied trophy in favour of a different narrative-approach to art making."

One of the featured artists, Claire Morgan, frequently includes in her art dead birds she has found in her neighborhood. Often, the birds and other objects, like bluebottle flies, bits of torn plastic, or dandelion seeds, are suspended on silk threads from the ceiling in her pieces. In several, the birds appear to be falling, crashing, or crumpling. The "unnaturalness" and precariousness of these art objects constitute a radical departure from traditional dioramas, which often reassure the viewer of the rightness of nature. Morgan's art, by contrast, unsettles that narrative and invites the viewer to witness things that are clearly out of balance.

The animal objects appearing in these pieces seem to fit more into the category of the souvenir than into the category of the trophy. They are souvenirs of the lost. Susan Stewart, in her book *On Longing*, writes of the desire surrounding a souvenir, "We do not need or desire souvenirs of events that are repeatable. Rather we need or desire souvenirs of events that are reportable, events whose materiality has escaped us." At a moment when we are facing what many are calling the sixth extinction, any animal body has the potential to be a souvenir. What's more, the sheer materiality of the dead animal is made more precious because it exists in this highly technological moment, in which we are flooded by the virtual and the digital. Perhaps Hatton's work reflects and embodies a desire to experience the unevenness of real life, an intimacy with real objects.

While Hatton is not skinning and stuffing honeybees, she does stage exhibitions of their dead bodies, and her work seems to arouse many of the same issues that Morgan's does. As noted above, Hatton's work also disrupts the tradition of the "natural" diorama we might find in a museum, thwarting the viewer's expectations by placing the honeybees in unlikely patterns—artistic forms, specifically mathematical ones. She began arranging the dead bees in sequences, like the Fibonacci sequence $(F_n = F_{n-1} + F_{-2})$, which can be found in the arrangement of seeds in the head of a

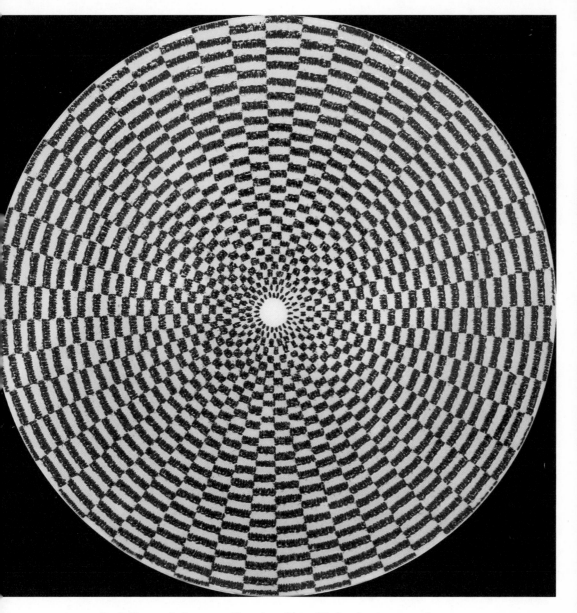

Sarah Hatton, *Circle 3*, 2014. Honeybees (*Apis mellifera*) and resin on panel.

sunflower. Knowing how fragile and small a bee carcass is, I was impressed by the careful, almost meditative process these pieces must have required. The thousands of bees were then coated with epoxy resin to preserve them.

The honeybees in Hatton's work become parts of the fixed patterns, emblematic of the structured lives they lead in this contemporary moment, bound by the rigidity of monoculture and commercial beekeeping practices. The mathematical patterns also create, Hatton hopes, a dizzying effect on the viewer, a visceral sense of the impact of pesticides on honeybees. The environment in which they are glued is a sterile white space, devoid of the diversity on which bees thrive.

The trapped insects become objects of mourning, but also objects of exhortation. Hatton herself describes the work as a balance of artistry and advocacy. Her images struck a chord with audiences internationally and were reproduced in magazines such as the *Huffington Post*, *Yes! Magazine*, and *Wired*. While her work does not directly affect current pesticide regulations, she has raised consciousness about the problem, and this may give individuals pause before they pour pesticides onto their lawns, and may even move them to pressure regulatory agencies to change their policies. While changing the science seems crucial, it may be that an interdisciplinary effort will be required to change policy and habit. Public opinion can help with that, and for this reason nonscientists are sometimes as important as scientists. Art can communicate in ways that science cannot. The endeavors of both scientists like Sai and artists like Sarah, working in concert, may be necessary to change our broken system and save the fragile and irreplaceable bees.

I realized that I needed to learn more about the role of farmers in the lives, and the fate, of bees. What imaginative innovations were occurring in the fields, on the farms where honeybees were crucial to production? My journey to discovering the answer to this question led me to some people who grow blueberries.

The Farmer, the Blueberry, and the Bee

AN UNUSUAL LOVE TRIANGLE

chapter five

As a young woman, I had a dream that I was very small and was flying fast through a forest of flowers, arcing in and out of vast cups of violet, through narrow crevices of vermilion, gliding past stamens and pistils, landing at last in the cathedral of an iris, the shrine littered with gold flecks of pollen. In the dream, I was not a bee, but a woman flying in a very small Wright Brothers–style airplane, hair blown back by the wind. When I woke up, I thought: what ecstasy to be a honeybee!

At the time, I had no idea that the visual rapture I imagined a bee must experience in her intimate relationship with flowers was just a portion of a much greater and even more sensual enterprise. Knowing what I do now, I can take the fantasy much further. Bees and flowers have quite a romance. I wasn't taking into account the electric charge on the blossom's edge, visible to a bee, the mingling scents of multitudes of flowers, or the whole-body experience of pollination. Nor had I thought about what it might mean to encounter a blossom covered with pesticides.

The exchange a bee has with flowers is a quite unusual interspecies relationship. It is no wonder that so many poets and writers have used the honeybee to represent sexuality. In "Spontaneous Me," Walt Whitman wrote:

Love-thoughts, love-juice, love-odor, love-yielding, love-climbers, and the climbing sap,
Arms and hands of love, lips of love, phallic thumb of love, breasts of love, bellies press'd
*   and glued together with love,*
Earth of chaste love, life that is only life after love,

The body of my love, the body of the woman I love, the body of the man, the body of
     the earth,
Soft forenoon airs that blow from the south-west,
The hairy wild-bee that murmurs and hankers up and down, that gripes the full-grown
     lady-flower, curves upon her with amorous firm legs, takes his will of her, and holds
     himself tremulous and tight till he is satisfied . . .

While Whitman's rather violent, masculine image of the bee is not accurate, his eroticization of pollination is understandable. The female honeybee flies into an open blossom, lands on the delicate satiny petals, and pushes her body between the erect stamens, tonguing the moist fronds until she reaches the perianth, where she can drink fully the aromatic nectar. As her body moves, the stamens vibrate and pollen spills from the anthers. The bee, whose body is coated with fine hairs, is covered with the dust of fertility, which she carries from flower to flower. This is what makes her an excellent pollinator.

In her lovely, intelligent book *Sweetness and Light*, Hattie Ellis traces the co-evolution of flowering plants and bees. Some scientists have looked at fossil records and believe that this relationship is 80 to 100 million years old. Without bees and other pollinators, like butterflies, flowering plants could not reproduce. Bees are responsible for pollinating at least one-third of the food eaten on our planet. As Michael Pollan suggests in *The Botany of Desire*, plants have learned to manipulate insects for their benefit. Plants' alluring colors, scents, and tastes seduce the bees into expanding their gene pool. And apparently this is good for the bees as well. Not only do they carry some of the protein-rich pollen home in their pollen baskets for food, but recent research in Australia suggests that bees actually have a pleasure hormone that is released both while they pollinate a flower and while they do their waggle dance upon returning to the hive. This gyrating dance guides other bees to the exact flower she was indulging in. Emily Dickinson famously, and apparently quite accurately, wrote, "To make a prairie it takes a clover and one bee, / One clover, and a bee, / And revery."

Not all flowers attract all bees. In fact, there are more than twenty-two thousand known types of bees on the planet. And different bees pollinate different flowers. The width and length of the opening and the scent and color of individual blossoms are designed to attract the best bee for the job. Charles Darwin gave a poignant example of this specific dependency in *The Origin of Species*, with a story about a certain variety of red clover that grew in a region of England. The red clover in question (*Trifolium pratense*) was pollinated solely by "humble-bees" (in the United States today, we know the humble-bee as the bumblebee). To illustrate the interconnectedness of the larger ecosystem, Darwin pointed out that mice tend to disturb the nests of humble-bees and eat them. "Now the number of mice is largely dependent, as everyone knows, on the number of cats," he explained. "Hence it is quite credible that the presence of a feline animal in large numbers in a district might determine, through the intervention first of mice and then of bees, the frequency of certain flowers in that district!"

Darwin's thesis was recently supported by a team of scientists in Colorado who discovered that certain plants were entirely dependent upon one species of insect. "Our results suggest that ongoing pollinator declines may have more serious negative implications for plant communities than is currently assumed," wrote Berry Brosi and Heather Briggs.

The reasons why certain bees choose certain flowers, and vice versa, have fascinated many entomologists and gardeners. What is it that attracts honeybees? Honeybees, or *Apis mellifera*, do not see the same color spectrum that we do. The scientist Karl von Frisch spent many years studying the sight of honeybees. In 1950, he published his book *Bees: Their Vision, Chemical Senses, and Language*, which described how his experiments proved that honeybees not only saw colors differently than we do but could also remember an exact color. The honeybee's eye is a stunning collection of tiny hexagonal lenses, as we can see in Rose-Lynn Fisher's images, that help the bee gather light. Morning meadows must vibrate with lambent color as she flies through them, using the sun as her compass. There is no red in her spectrum, so lavenders, yellows, and blues would be the most radiant colors. As we know, honeybees are interested in the flowers of many fruits, vegetables, and nuts, but many flowers require different pollinators. Lupine is most attractive to long-tongued bees like bumblebees and some mason bees. Joe-Pye weed also attracts the bumblebee. False indigo is thought to be visited primarily by the mining bee, but also by leafcutter and polyester bees. The list goes on and on. It does seem fairly egocentric to think that the human is responsible for the propagation of flowering plants. In his essay "Border Whores," Michael Pollan meditated on this idea:

In the space of this garden, I tell myself, I alone determine which species will thrive and which will disappear. I'm in charge here, in other words.

But that afternoon in the garden I found myself thinking that a bumblebee would probably also regard himself as the decision maker, opting for one bloom or another. But we know that this is just a failure of his imagination. The truth of the matter is that the flower has cleverly manipulated the bee into hauling its pollen from blossom to blossom.

I love Michael Pollan's point. Insects do indeed ensure the survival of the plants that have evolved to collaborate with them. Unfortunately, however, humans have intervened in this lovely symbiotic relationship by using pesticides and GMOs and have thus completely changed the game. And this is happening on a large scale owing to our dependence on monocrops.

Monocrops radically diminish the biodiversity of any agricultural area. I thought again about Liittschwager's cornfield, devoid of life except corn, a single mite, an ant, several grasshoppers, and a small mushroom. The intensely sensitive body of the honeybee is imperiled by our current agricultural practices, which depend on both bees and, unfortunately, pesticides. Because bees are responsible for pollinating so many food crops, they are obviously important to farmers. What are farmers to do? To me, the answer seems obvious. If you need bees, quit using pesticides. But I learned that the choice is not quite so easy to make. Still, there are people who are rethinking our system.

I knew from a young age that pesticides could be very detrimental to human and nonhuman beings. I was the child of parents born in 1944, who lived through the publication of *Silent Spring* and the ripples of discussion and activism it sent forth throughout the nation. And I had heard stories. Spending much of my time as a little girl in a small town in northern Illinois, I had heard the legends about John Eisbach. My mother told us with sober admiration and sadness how he had been an environmentalist long before environmentalism was an established movement. He was an organic blueberry and tomato farmer whose death was blamed on pesticide poisoning. His case had been defended by a Mayo Clinic hematology expert. His son, John Eisbach Jr., and his daughter, Christina, still lived on his property, Wooded Wonderland. I decided to visit them and see how their blueberries and bees were faring.

John Jr. looks like a stereotypical lumberjack: broad shoulders, chiseled bronze face, shrewd eyes always narrowed in concentration. His hands are thick and knotted, like limbs of an ancient oak. One bears the evidence of a tussle with a saw—not quite disfiguring his hand but reshaping it into proof of his tenacity. Christina is tall, strong, and clear-eyed—a younger, female version of her father. They met me on the dirt trail that runs through the bottom of their valley, not far from John's sawmill. The property has areas of prairie and forest, as well as several acres of blueberries and smaller vegetable gardens. John's voice is deep and soft, and his words come slowly and thoughtfully. I knew I was in the presence of a woodland sage.

John's knowledge of the land, and of the trees, animals, and insects that inhabit it, has drawn the attention of scientists, naturalists, and locals interested in learning how to do what he does. His lumber business is completely sustainable. He doesn't use any pesticides on his blueberries. His healthy, ruddy-cheeked grandchildren are helping around the farm. He's doing something right.

John's blueberry fields, a congregation of bushes eight to ten feet tall, startled me, not because of their size and shape but because they flickered and blinked. All of the trees appeared to be covered with tinsel and ready for some kind of celebration. I learned that John and his helpers had tied hundreds of feet of discarded VCR tape into the trees to create this dazzling effect. Here was one of John's secrets of pest management. The shiny trees kept blue jays and deer away. A shrill shriek pierced the air. The recorded calls of birds of prey came periodically from a loudspeaker to keep other small berry eaters away. The soil here is naturally acidic, which blueberries like, though John periodically adds oak sawdust. The sawdust rots, providing nutrients and acidity to the soil, and also serves as mulch. John prunes the trees himself. Not far from the blueberries are about twenty-five beehives. "The bees have plenty to eat all year," John told me, "starting with basswood, an early bloomer. No bees, no fruit. Simple as that."

John's father planted all of these trees, and he built the roads and the earliest buildings. Tears fill John's eyes as he tells me about his father's tomato gardens, hoop houses, and the first blueberry bushes, about how he taught John everything he knew starting when John was five years old, and about his untimely death from the pesticide that destroyed his blood. The fight against the chemical company that sprayed and destroyed his father's body still haunts John. No monetary compensation could ever adequately replace a life. And the compensation was not large. In a chemically dependent society plagued by risk at nearly every juncture, it is hard to pin blame. John's response is to live close to the ideals he hopes the whole society will begin to take on.

"Scale is one issue," John said, squinting back his emotion. "Scale and greed. Things change when you are shipping fruit and bees all over the globe." John believes that local food is one answer. His own customers are aware that not all the blueberries will look alike. He sells his blueberries and his honey to local restaurants. And he doesn't need to sell as many berries as big growers do. Small scale was working for him. But, I wondered, was it really inevitable that large-scale monocrop farms had to be so destructive to the environment?

To find an answer to this question, I sought out Dr. Rufus Isaacs, an entomologist and pollination architect working in Lansing, Michigan. Isaacs has been working with blueberry farmers to explore the effect on fruit yields of planting native wildflowers on the edges of their crops. Was it possible to have healthy

89

pollinators and high productivity for large-scale farmers, too? I packed my bags and headed to Michigan.

———————

Highway 31 South curls along the shore of the western edge of Michigan. In mid-July, everything was verdant. Coming around a corner, neat lines of cherry trees articulated the curvature of the low hills. Around another bend, a lush vineyard seemed to be waving its palm-sized leaves at the passing cars. Now and then, I passed through a small town flanked by meadows mottled with the white sprinkle of Queen Anne's lace and the ethereal lavender of chicory, and then suddenly I was in a stand of pine with the brilliant blue of Lake Michigan gleaming through the passing trees.

Part of the reason for the lush roadside vegetation was the support of the USDA's Conservation Reserve Program. Funding incentives encourage landowners to create environments that support wildlife. Traditionally, bird enthusiasts have supported the program's protection of wetlands and grasslands, but more recently, pollinator conservation groups were standing up for fields of wildflowers. The USDA Natural Resources Conservation Service was funding pollinator-friendly plantings, which are heavy on wildflowers and light on grass. The result was a feast for the eyes. It was certainly nothing like Liittschwager's cornfield.

Michigan State University professor of entomology Rufus Isaacs met me on the southeastern side of the state in South Haven, Michigan, home of the annual blueberry festival, which features blueberry-pie-eating contests, blueberry pancake breakfasts, a tractor pull, live music, and a blueberry parade. Rufus's gentle British accent and kind eyes made me feel welcome at once. Rufus grew up in England and spent his free time on a sailboat. He had originally planned to become a marine biologist, but in his junior year in college, an internship sparked his interest in insects, and he became an entomologist. After finishing his degree, he and his wife moved to the United States. His role as an extension professor at MSU allows him to do his own research and also to work with a community of farmers. The first thing on Rufus's agenda was to introduce me to one of the blueberry growers he had been working with for years, Dennis Hartmann. We drove to the designated meeting place, the local bowling alley.

Dennis stepped out of his huge white pickup truck, balancing a couple of five-pound boxes of blueberries in one hand. He gave us a dimpled grin from under his shock of blond hair and handed each of us a box of freshly picked berries, insisting that I try them. Every one of the dark blue berries in the box was about the diameter of a nickel. The explosion of sweet, cold juice on my tongue was as refreshing as a dip in Lake Superior. This flavor was not the only selling point, either. The berry had what Rufus described as "the blue halo," thanks to its antioxidant content and memory-enhancing benefits. A handful a day, Dennis recommended. With a warm smile like Dennis's and blueberries this good, I could understand why Dennis was a huge success.

Dennis and his wife, Shelly, own True Blue Farms, which packs and sells approximately 4.5 million pounds of blueberries a year to grocery stores and baking companies like Sara Lee; some travel as far as New Zealand. They sell their own products as well. At their retail outlet, called the Blueberry Store, you can buy blueberry syrup, blueberry jam, blueberry muffin mix, blueberry ice cream—even blueberry beef jerky. It's an enormous enterprise. But True Blue wasn't always this big. The Hartmann family started with just ten acres.

Dennis's grandfather and a handful of other farmers pioneered the Michigan blueberry industry. By the 1940s, most of the southwestern coast of Michigan had been cleared by loggers, and the acidic soil left behind was perfect for growing blueberries. With the encouragement of Stanley Johnston, Dennis's grandfather and his neighbors began planting bushes. Johnston grafted the highbush and lowbush blueberry together to create a plant that would grow well in southern Michigan. He thrust armfuls of cuttings into the arms of residents and told them to throw them in the ground. Today the entire region is covered with fruit orchards, and all of those orchards need pollination.

"We love bees," Dennis said. The pollination of blueberries is crucial to his business, and he depends on beekeepers to bring bees to his orchards in the springtime. The window for pollinating a blueberry blossom is only about four days. Dennis calls his "bee man" when he has about 20 percent bloom. Each acre of blueberries needs four or five hives. The blossom of a blueberry is shaped like a little bell, a tiny, sweet-smelling room for a bee to explore. While the honeybee is an excellent pollinator of these delicate reproductive organs, the bumblebee is even more efficient. Like Whitman's wild bee, the bumblebee holds herself "tremulous and tight till [s]he is satisfied." The bumblebee is an enormously efficient pollinator because she can detach the muscles used to move her wings and use all of her energy to vibrate her body—"like an electric toothbrush," Rufus tells me—and instantaneously cover herself with pollen. While a blueberry blossom needs to be visited by several honeybees to complete the pollination, a bumblebee can do it in one stop. The bumblebee might seem like the wiser choice for a farmer interested in pollinating a crop, but in fact bumblebees live in small communities and are not as easy to keep. To Dennis and Rufus, the best scenario is to have both kinds of bees, and to support other pollinators as well. The next day, Dennis would give us a tour of the fields themselves.

Rufus pulled his car onto a little gravel road that led into the Trevor Nichols Research Station. Behind a handful of small buildings used by Michigan State University, 120 acres of fruit orchards stretched over the rolling hills. We passed trees heavy with pears and covered with bright red cherries, and rows of grapevines, finally entering a field lined with blueberry bushes. Rufus introduced me to the varieties that are chosen for their unique seasonal rhythms as well as for their taste. The hope is to have blueberries coming to maturity for as many weeks as possible. We had arrived when several types were at the height of harvest. The Jersey and the Bluecrop were the most popular for eating by the handful—sweet and juicy and plump. The Rubel had a bit more tanginess, which made it good for pies. They all tasted good to me.

A couple of the blueberry bushes were sporting what looked like little gauze mittens. Early in the spring, when the bushes first came into bloom, Rufus and his grad students chose a random selection of nice-looking blossoms and enclosed them in these "mittens," which allowed sunlight and rain in but kept pollinators out. The experiment aimed to determine how the fruit would change if pollinators did not visit it. Rufus opened one of the small sacks and retrieved a few berries. The berries were substantially smaller than those on the mitten-free parts of the bush. In his other hand he picked a few that had been open to insect pollination. Rufus pulled out a credit card and carefully sliced a couple of berries in half. To be honest, I had never thought about the seeds in a blueberry. They are too small to really notice when you're eating them, but when Rufus opened up the berries, the seeds were obvious. A perfect circle of approximately thirty tiny brown seeds sat in the center of the pollinated berry. In the other blueberry there were only a few. Not only did the insect pollination ensure larger and more plentiful fruit; it also ensured better fertility in the plants themselves.

Next, we drove to a blueberry orchard owned by the Hartmanns. Flanking the crop was a field of native wildflowers that Rufus had designed specifically for Dennis. A host of deep pink coneflowers, light pink milkweed blossoms, and the bright yellow stars of the cup plant danced in the breeze. This collection of flowers, which also included showy goldenrod, wild lupine, sand coreopsis, and New England aster, was a feast for a wide range of pollinators: butterflies, blue orchard and other mason bees, cellophane bees, bumblebees, and, of course, honeybees. As a honeybee nuzzled a blossom, Rufus looked on, beaming. This field was good for flowers, good for farmers, and good for bees.

But blueberry farmers have competition. Many insects like to eat blueberries as much as humans do, and they are not only interested in the berry itself. Blueberry buds are the favorite of blueberry bud mites, cutworms, and spanworms. The delicate bell-shaped blossoms are a delicacy for blueberry blossom weevils and flower thrips. The leaves and shoots are a popular item among aphids, blueberry flea beetles, blueberry gall wasps, blueberry tip borers, gypsy moths, oblique-banded leafrollers, scales, sharp-nosed leafhoppers, and white-marked tussock moths. The roots are tastiest to blueberry mealybugs and Oriental beetles. And the human favorite, the blue fruit itself, is loved by blueberry maggots, cherry fruitworms, cranberry fruitworms, plum curculios, spotted wing drosophila, and three-lined flower beetles. These are clever insects. The obliquebanded leafroller, for example, not only weaves little webs between the leaves for its own protection, but it also rolls itself into a delicate leaf sleeping bag, which serves as marvelous camouflage. The gypsy moth mother, from a family of infamous defoliators, builds her silken nest in nearby trees, and when the larvae are old enough, they rappel down from the highest branches and dangle from a silk until a good wind comes by and lofts their adventurous bodies up into the air and carries them to their destination—a blueberry bush, if they are lucky! The scale, to take another example, builds itself a tiny tent on the berry itself, under which it stays and drinks sweet blueberry juice to its heart's content. When the insect moves on, it leaves a tiny green spot on the berry. Evidence like this spot, or residual frass, or silk webbing, or any other imperfection is not tolerated by most blueberry consumers, and the farmer has trouble selling imperfect fruit. What is a grower to do?

In 1941, Row, Peterson and Company produced a textbook for their Basic Science Education Series called *Insect Friends and Enemies*. The book, written by Bertha Morris Parker and zoologist Robert Gregg from the University of Chicago, acknowledges that some insects, like pollinators, are our friends, but emphasizes that the vast majority are against us. Section titles like "Insect Enemies of Our Orchards," "Insect Enemies Inside Our Buildings," and "Causers and Carriers of Disease" are accompanied by illustrations of insects and their destruction: a worm-infested apple or a leaf turned to lace by beetles. At the end of the book, the authors explain in simple language, "As you have already been told, insects are our greatest rivals. Some people believe that in spite of all we can do our insect enemies will bring about the end of civilization. Some people even believe that the insect will at last crowd man off the earth altogether. With what you have learned about our insect friends and enemies, you are left to answer the question for yourself. Will the insects win?"

Indeed, this attitude seems appropriate when you look at the insect populations recorded in the early part of the twentieth century. In *Insectopedia* (2010), Hugh Raffles tells the story of some scientists in Louisiana who first began measuring insect samples with airplanes. The pilots flew at different altitudes, gathering insects at different times of day and night and in all kinds of weather. "They estimated that at any given time on any given day throughout the year, the air column rising from 50 to 14,000 feet above one square mile of Louisiana countryside contained an average of 25 million insects and perhaps as many as 36 million."

Following World War II, pesticide use exploded. Many saw the introduction of DDT as a godsend. Some insects carried diseases that were deadly to humans. Mosquito-borne diseases like malaria were taking thousands of lives every year. From that perspective, DDT was a savior chemical. In a fifteen-minute documentary called "DDT—Weapon Against Disease," produced by the U.S. War Department in 1945, the lauded miracle chemical was shown being blown into swamps, sprayed into kitchens, and even sprinkled in powder form into the hair and over the bodies of children in malaria-prone countries.

In addition to their ability to eliminate disease-carrying pests, pesticides were seen as a boon for industrializing farmers. In *Paradox of Plenty: A Social History of Eating in Modern America*, Harvey Levenstein discusses the American push to eradicate world hunger in the early 1960s, when industrial food was enjoying a heyday. If you could produce more food faster, fortify it with vitamins, and package it for easy shipping, you could save millions of starving children. This well-intentioned push for bigger and faster production was augmented by the use of pesticides that could ensure fewer bad bugs and higher yields. What was not yet fully understood was the impact on soil, water, pollinators, and humans of the rampant application of these chemicals.

John Eisbach's customers understand that organic produce may have aesthetic blemishes, and they do not demand the same kind of perfection that True Blue's do. Dennis Hartmann explained that when he sends a shipping container full of frozen blueberries to New Zealand, the container can be rejected if even one Japanese beetle is found among the millions of blueberries. True Blue Farms delivers what the customer wants. The Hartmanns' huge team works to ensure that the blueberries are perfect. They even have a long machine, about the length of a city bus, that shakes small twigs free on trembling wire surfaces, gently washes the berries, and even checks them for color. The berries travel along on a little conveyor belt and enter a photo booth the size of an oven. As they pass through the space, a color scanner identifies unripe blueberries. The blue berries are thrust over a gap with a burst of air, but detection of a green berry reverses the direction of the air and sends the bad berry down to a bin below. The berries are then inspected by human eyes before they are shipped. No imperfect berry can sneak through this operation.

What if our expectations changed? What if we were willing to eat imperfect fruit? But what about the issue of blueberry pests? Again, Rufus Isaacs has an idea, which echoes the work of Rachel Carson. What if other plants—plants that attract beneficial insects who would eat the blueberry predators—were planted next to the blueberry bushes? This would reduce the need for pesticides. And this is exactly what Rufus was doing in Michigan with the Integrated Crop Pollination Project.

The fields of wildflowers that Rufus planted next to the Hartmanns' blueberries not only provide good food for pollinators; they also attract "beneficial insects" who eat blueberry pests. Certain native flowers encourage populations of lacewings, minute pirate bugs, hoverflies, damsel bugs, and ladybugs. Even certain wasps are considered helpful. These plantings yield results after two to three years, Rufus explained—and they work! Rufus and his colleagues at Oregon State University, Loyola University Chicago, Rutgers University, Pennsylvania State University, the Xerces Society, and several other organizations have jointly received funding to extend this kind of planting experiment to more crops all over the United States. With the Integrated Crop Pollination Project team, growers of apples, cherries, blueberries, watermelons, raspberries, pumpkins, and almonds will be planting new crops next to their cash crops. These crops will be healthier for the land and the insects, and should prove beneficial to the farmer's pocketbook as well.

Perhaps slowly, with a shift in consumer conceptions of perfect fruit and changes in growing practices, the lives of bees, flowers, and humans will improve. After all, we are all in this together.

But what if things do not change, and bees and other pollinators continue to disappear? Without pollinators, we might need to think of other solutions. One possibility is mechanical bees.

Elizabeth Goluch, *Bumblebee.* Sterling silver, 14K and 18K gold, 7 × 7 × 4 in.
Collection of David Weishuhn, Toronto, Canada.

Techno Bees
The Art of Elizabeth Goluch

Canadian metalworker Elizabeth Goluch grew up on a farm in northern Ontario, where her interest in the natural world was cultivated. She received her BFA from the Nova Scotia College of Art and Design in Halifax, Nova Scotia, in 1976. For the past ten years or so, she has been making one-of-a-kind, larger-than-life insect sculptures out of gold, silver, enamel, and gemstones. The pieces reflect a keen understanding of the anatomy of individual insects. These metal insects have global appeal and have been shown in galleries and private collections in the United States, the Czech Republic, Denmark, Australia, and Hong Kong. Goluch says of her work, "The decorative details and the treasure contained in each piece reference elements of insect life, lore, and environment." Many of her complicated sculptures have secret chambers that the viewer can open to discover more features. Nestled behind tiny doors are rich treasures: smaller sculptures of anything from a flower with a gemstone center to a musical note made of gold. Assembled out of highly valued metals and gems, Goluch's artworks honor the strangeness,

the complexity, and the uniqueness of insects, but also their precious irreplaceability. The creations are visually striking and quite beautiful, but the prominent rivets and segments are also reminiscent of the steampunk aesthetic and suggest a certain toughness in a postapocalyptic world. Bees made out of metal may seem strange at this historical moment, but in some visions of the near future, they may become quite commonplace in the form of the RoboBee, an object designed for very different purposes from the sculptures of Elizabeth Goluch.

Perhaps born of a similar fascination with the amazing design and capabilities of insects, engineers at Harvard's School of Engineering and Applied Sciences are collaborating with faculty from the Department of Organismic and Evolutionary Biology at Harvard and Northwestern University's Department of Biology, a microelectronics firm in Washington, D.C., and the Wyss Institute for Biologically Inspired Engineering (also at Harvard) to create a mechanical bee that will mimic many of the real qualities of the honeybee. The tiny robotic bees, called RoboBees, will not only

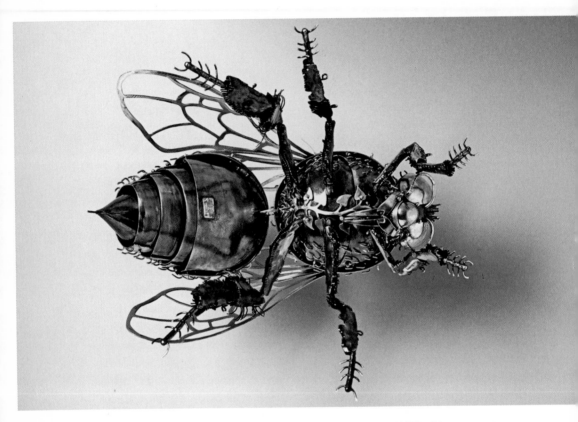

Elizabeth Goluch, *Bumblebee,* underside. Sterling silver, 14K and 18K gold, 7 × 7 × 4 in.
Collection of David Weishuhn, Toronto, Canada.

be able to "sense" other robots and natural flowers and other plants; they will also be programmed to perform in concert, the way a real community of bees would behave, as a colony. The idea is to "simulate the ways groups of real bees rely upon one another to scout, forage, and plan." The planning and decision-making qualities of these robots make them unique. The prototype successfully flew, hovered, and landed in 2007, and RoboBees are already being used in some medical applications. The team at Harvard believes that crop pollination could become feasible within twenty years, although they do not see RoboBees as a viable replacement for real honeybees. The team makes clear on its website that "we do not see robotic pollination as a wise or viable long-term solution to Colony Collapse Disorder (CCD). One of the potential applications of micro-robots might someday be to artificially pollinate crops. However, we are at least 20 years away from that possibility. Furthermore, even if robots were able to be used for pollination, it would only be as a stop-gap measure while a solution to CCD is implemented to restore natural pollinators." Additional uses for the tiny, coordinated teams of robots would include search and rescue, hazardous environment exploration, weather and climate mapping, traffic monitoring, and military surveillance.

While I deeply admire the innovative thinking that goes into creating a robotic bee that can sense things and communicate with other RoboBees, I wonder about the impact on birds, who might mistake them for food in the same way that sea turtles often mistake plastic bags for jellyfish, or the impact on other pollinators, who might be in competition with these robots. A poem called "One Day," by Laurie Sheck, comes to mind, in which, years from now, creatures from another planet visit the earth and find only our technology:

One day long after we are dead
the strangers will come
to look down at us
from their great distance,
the blue planet like a coin
from some ancient currency
stamped with a language blurred as ruined film.
How quiet it will be then,
the cities deserted except for the machines
still crawling over the soil
like dreams broken free of their skulls.
What did we want of the earth?
What did we ask of it? We turned
our faces away. Will they learn of us,
studying the few found texts,
the fragments of enterprise and greed,
how we disguised our fragility with stone,
how we worshipped power and feared
the softness of our own bodies,
how we crisscrossed the earth with our wires
as if the constant play of voices in the air
could make us listen. But the trees
grew hushed, the waters more distant. . . .

Will RoboBees outlast us, I wonder, like the machines in Sheck's poem?

Somehow, Elizabeth Goluch's metal bumblebee does not give me the same impression. What strikes me are the mysterious complexities of the real insect that

Elizabeth Goluch, *Bumblebee*, overview (*above*) and thorax (*opposite*). Sterling silver, 14K and 18K gold, 7 × 7 × 4 in. Collection of David Weishuhn, Toronto, Canada.

her representation honors. Somehow, this human-made insect sublimates the desire for domination or replacement through mimicry and instead invites the viewer to feel a sense of wonder about a being that is radically different from the human. Goluch's models include earwigs, cicadas, pill bugs, dragonflies, mud dauber wasps, even cockroaches. She celebrates an insect biodiversity no longer common in our society. Most of our energy is spent killing off these insects to keep our spaces insect-free, especially sacred spaces like lawns.

For the Love of Lawns

Springtime in the Midwest is a time of euphoria. The days and nights of bitter cold, with winds capable of breaking everything from branches to spirits, are finally over. When the thaw arrives, everyone is outside collecting evidence that winter is truly finished, delighting in the sound of the newly formed rivulets and dripping icicles, as even the movement of water seems like a miracle. Soon the tops of the maple trees burst into bloom. Crocuses emerge, then blue squill and daffodils. Birds return, and bees begin their cleansing flights on warm days, and as soon as those flowers are ready, they begin to forage.

This past spring, like the bees, I was energized by the sunlight and warm air. One sunny day, I pulled on my running shoes and coat and set out for a jog. I headed toward the nearby restored prairie and arboretum by cutting through the neighborhoods near my house, neighborhoods filled with single-family middle-class homes. Although I struggled to get my sedentary winter body in the mood to move again, I was thrilled to be breathing fresh air in big gulps. I was glad to be outside, in "nature" again. But I was soon reminded that this fantasy about getting out into the healthy, clean spring air was foolhardy. The air, which in my spring fantasy should be filled with the scent of apple blossoms and freshly cut grass, was in reality perfumed by the strangely sweet smell of lawn chemicals. Soon I spied the small red and white flags that warned humans, at least: "Keep off. Pesticide Application." I passed lawn-care workers with tanks strapped to their backs administering chemicals in a fine spray through a hose. I heard the first mower of the season roar to life. So began the spring ritual of attaining the perfect lawn.

Americans love lawns. To get a sense of just how much, I turned first to a study conducted by NASA's Cristina Milesi, who, along with her team, employed satellite and aerial imagery to quantify the amount of lawn space we have in the United

States. Their most modest estimate was approximately 163,812 square kilometers. To translate their estimate, that is approximately 40 million acres of lawn.

How big was 40 million acres of lawn, anyway? "Even conservatively," Milesi said, "I estimate there are three times more acres of lawns in the U.S. than irrigated corn." Did this compare to any other crop? To get a better sense of this quantity, I went to the National Junior Horticultural Association website, where I thought I might find some very basic figures. And indeed I did. The USDA produced a report on how many acres of different vegetables were grown in the United States in 2015, and it is interesting to note that the featured acreages have been shrinking over the past few years. The acreages of the most common fresh market vegetables in 2015 were:

Beans, snap—77,680 acres   Cucumbers—40,020 acres
Tomatoes—97,500 acres   Onions—143,900 acres
Peppers, bell—44,800 acres   Squash—41,250 acres
Cabbage—59,530 acres

I compared these figures to those 40 million acres of lawn. It was a lot of grass. To make this even clearer, in his book *Lawn People*, Paul Robbins writes, "Covering a total area roughly the size of the state of Iowa, the lawn is one of the largest and fastest growing landscapes in the United States. The lawn also receives more care, time and attention than any other space." But what does that care look like?

Luscious thick green turf depends on water, so most lawn owners dutifully connect their hoses to their sprinklers and give their grass as much water as it likes. Milesi's team estimated that "about 200 gallons of fresh, usually drinking-quality water per person per day would be required to keep up our nation's lawn surface area." Some estimates suggest that 30 to 60 percent of freshwater is pumped onto lawns, depending on the city. Freshwater is a precious commodity, and as the earth warms and droughts become more frequent, it will become even more precious. The severe long-term drought in California has motivated Californians to be more conservative in their water use, but most of us don't think about this issue. We water our grass until it grows thick and tall, and then we quickly cut it down again.

No one wants a shaggy, spotty, weedy lawn. That kind of lawn suggests neglect, sloppiness. A good lawn must be mowed to keep that turf looking like a plush carpet. A 2008 article in *Scientific American* claimed that we have "54 million Americans mowing their lawns on a weekly basis," and that "gas lawnmower emissions account for as much as five percent of the nation's total air pollution." The

Electric Power Research Institute reports that if Americans replaced half of the 1.3 million or so gas mowers in the United States with electric models, it would be equivalent in terms of emissions to taking two million cars off the road. That's a lot of fossil fuel use. But that's not the only way we consume fossil fuels for lawn care. When the lawn looks thirsty, we water it. But we also feed it, with fertilizers and pesticides.

The National Gardening Association boasted that a hearty $29.5 billion was spent on lawns and gardens by U.S. households in 2013. A good portion of that money pays for pesticides. "Risks from Lawn Care Pesticides," a research paper generated for the nonprofit Environment and Human Health, reported that 80 million pounds of the active ingredients in pesticides are used on U.S. lawns annually. If private homeowners use 80 million pounds of pesticides per year to keep our lawns looking good, how does that compare to farmers? Robbins says, "Lawn pesticides are applied on a scale to rival agricultural toxins; 23% of the total 2, 4-D applied in the United States in used on lawns; 22% of glyphosate, 31% of chlorpyrifos, and 38% of dicamba." In 2000, the U.S. Fish and Wildlife Service claimed that household use was ten times that of farms.

At the National Pesticide Information Center, one can find fact sheets on these chemicals that we're liberally applying to the lawns on which we, our children, and our pets play, and in which our insects and birds try to survive. Dicamba, it appears, is one of the chemicals we Americans use most plentifully. Dicamba toxicity in animals is described this way: "Signs of dicamba-induced toxicosis in animals include shortness of breath, muscle spasms, cyanosis, urinary incontinence and collapse. Additionally, salivation with tympanism (excessive gas in the gastrointestinal tract) along with possible depression and convulsions may occur. . . . As a result of a single oral dose of DMA dicamba, bobwhite quail exhibited symptoms such as wing droop, loss of coordination, weakness, abnormal gait, rigidity of the legs and lethargy." Chlorpyrifos, which is used on golf courses, is a broad-spectrum insecticide designed to destroy the nervous system by inhibiting the breakdown of certain neurotransmitters, eventually causing death. This chemical is only licensed for *indoor* residential use inside containment boxes because it is too dangerous to apply directly to residential lawns. "Signs from intermediate syndrome," according to the National Pesticide Information Center, include "weakness of the neck, front limbs and respiratory muscles, diminished appetite, depression, diarrhea, muscle tremors, unusual posturing and behavior (including cervical ventroflexion), and death." The NPIC fact sheet continues, "Data from two human studies indicate that humans may be more sensitive to chlorpyrifos compared to rats or dogs, following acute oral and dermal exposure." Obviously, these chemicals are not only bad for insects. Perhaps if they were only used in tiny amounts in emergency situations (though what that emergency might be, I can't imagine), it might seem justifiable, but 80 million pounds seems terribly dangerous.

In *Redesigning the American Lawn: A Search for Environmental Harmony* (first published in 1993), F. Herbert Bormann, Diana Balmori, and Gordon T. Geballe explain that although Rachel Carson successfully sounded the alarm about DDT, pesticide use had actually doubled since her book was published in 1962. They quote an EPA estimate suggesting that even back in 1984, "more synthetic fertilizers were applied annually to American lawns than the entire country of India applied to all its food crops." The amount of fossil fuel being burned for the sake of maintaining millions of acres of mowable turf struck me as obscene.

It had seemed so easy to point the finger at industrial agriculture for causing the pollinator crisis, but in fact the administration of pesticides and herbicides for agricultural use is highly regulated. In 1990, California became the first state to require monthly pesticide-use reports for all agricultural purposes. The state's definition of agricultural use includes "parks, golf courses, cemeteries, rangeland, pastures, and along roadside and railroad rights-of-way." In Wisconsin, growers must obtain licenses to apply pesticides, and they often work closely with the state's Department of Agriculture. Wisconsin has a "Drift Watch" program that allows growers to look at maps of cropland when they are planning their spraying (and its timing) in order to prevent the contamination of organic crops or bees by drifting pesticide. While the safety standards set by the government are clearly not perfect, many growers work to stay within those limits, whereas most residential pesticide users don't have to adhere to any rules. They can choose to follow the directions on the bottle of Roundup or not. And all for the sake of perfect lawns.

Where did the American obsession with lawns come from? Why have we chosen this particular landscape aesthetic? Bormann, Balmori, and Geballe trace the origins of the lawn to the early colonial period. The colonists brought with them the landscape aesthetic of Great Britain, a country covered by large tracts of pastureland. Upon their arrival in the "New World," colonists were alarmed by the paucity of grazeable land, and they quickly cut down trees and began planting European timothy, bluegrass, and white clover. "Within one or two generations, these plants had become so common that settlers regarded them as native." Thomas Jefferson is said to have greatly admired a certain building in England surrounded by "the lawn, about 30 acres." He re-created his dream lawn at his own estate, Monticello, where he employed deer to help with the work of keeping his grass short. Mown grass seemed to be a symbol of having attained control over "wild" nature, a measure of the colonists' success.

Interestingly, it was this same impetus to create a successful colony that brought honeybees across the Atlantic in the first place. In *Bees in America*, Tammy Horn describes how, in the pursuit of a biblical "land which flows of milk and honey," colonists transported honeybees to the colonies to produce both honey and wax for candles. It was not a simple transition, because survival was difficult for the humans and the bees, but eventually the bees adjusted to their new surroundings. "The bees have generally extended themselves into the country, a little in advance of the white settlers. The Indians therefore call them the white man's fly, and consider their approach as indicating the approach of the settlements of the whites," wrote Jefferson in 1785. Perhaps it is not altogether ironic that the drive for success and domination in American culture, symbolized by the American lawn, would eventually destroy the very bees who once stood for these same impulses.

The real expansion of the American lawn followed the birth of the suburb in the mid-nineteenth century. That and Edwin Budding's invention of the lawnmower in 1830 made possible today's chemically laden expanses of green, decorated with lawn furniture, swing sets, and golf carts, but very little clover. Our addiction to chemically sterilized landscapes is sucking the sweetness out of the land of plenty.

In front of a few houses in my neighborhood, there are bright explosions of yellow sprinkled throughout the lawn. These are the dandelion yards. The dandelion is a favorite of honeybees, so I let the weed bloom all over my yard and delight in seeing my bees pushing their way through their feathery tops. I am always tempted to knock on the doors of the other dandelion-yard homes and thank them for not using chemicals, and tell them that I truly love dandelions, from their bright yellow youth to their fluffy gray maturity. My next-door neighbor, however, hates them, and because he is a very environmentally conscious person, every spring he crawls around his yard on his hands and knees with a special tool that looks like a screwdriver with a V-shaped nick at the end of the blade and patiently digs them out, one by one. And because we care about each other deeply, he does not complain about my dandelions, which must reseed his yard year after year, nor do I beg him to leave them alone so the bees can sip their nectar and bring the bright orange pollen back to the hive. Instead, he drops the pile of dead weeds over my fence, and I feed them to my chickens.

Not everyone has such a neighborly arrangement, however. Lawns can be a very touchy business. I remember, as a child, that a friend of my mother's named Tanny Hilbert was sued by a neighbor for letting her yard go wild with weeds. Knowing Tanny, I'm sure she felt very good about feeding the butterflies and bees, and about giving the rabbits and birds a place to nibble or seek shelter. The yard was always full of flowers and grasses of differing heights, lovely to some, hideous to others. Tanny lost the case and was ordered to mow her lawn. Many community associations today have very strict rules about what you can and cannot grow in your yard. Why do lawns cause so much tension?

In *Lawn People*, Paul Robbins suggests that the lawn has become a symbol of the collective good. "Such a lawn only developed as a product of economic growth conditions in suburban real estate developments," he writes, "tied to proselytizing that connected the lawn with a certain kind of desirable urban citizen and economic subject. . . . The management of this collective good is [experienced] not so much as a choice, but as an obligation, which brings with it anxieties and mixed feelings . . . the nagging sense that something might be wrong." Property values, municipal codes, and homeowners' associations keep even environmentally conscious people from doing away with intensive-chemical lawn care. "And more people are becoming lawn people all the time," says Robbins. So there is certainly pressure to keep the perfect green lawn going.

As I wound through the neighborhoods last spring, though, I did see something that felt heartening. People's yards were not only a sea of green. I did see tiny spears of hydrangea, tulips, and hosta lilies, and in some places patches of bright blue. Many homeowners were choosing flowers over all grass, and many people were out buying more. Spring is the time when not only is everyone outside but every big box store is selling flowers to beautify our yards. Pots full of geraniums, pansies, and lilies are loaded into Subaru hatchbacks and taken home to plant. The sight of flowers made me happy, especially because I had a hive of bees behind my own house, not too far away, and I was pleased that there would be blossoms for them to find. This bit of solace was undermined, however, when I read a study about how most commercially grown garden plants contain systemic pesticides, including, very frequently, neonicotinoids.

A study released in June 2014 by Friends of the Earth and the Pesticide Research Institute, a collective including Timothy Brown and Susan Kegley, showed that 51 percent of the pollinator-friendly plants sold in stores like Lowe's, Home Depot, and Walmart contain dangerous quantities of systemic neonicotinoids, which means that the very plants people were planting to help pollinators survive were in fact killing them.

This news was bad enough on its own. But then I thought about the fact that 30 to 60 percent of an area's freshwater is used in watering lawns. I thought about all the pesticides being poured onto the lawns directly, plus all the pesticide residue running off into our streams, lakes, rivers, and ponds. In a study published in *PLOS ONE* in May 2014, Anders Huseth and Russell Groves reveal that neonicotinoids not only affect insects when the plant is alive but continue to leach into the soil and groundwater after it has decomposed. In a survey of one agricultural region in Wisconsin, they discovered that the leaching of the pesticide was strong enough that it could be detected in the water coming out of the sprinkler pivots that water the fields. The concentration was high enough to have an adverse effect on honeybees. The study concluded that we are essentially "recycling . . . neonicotinoid insecticides from contaminated groundwater back onto the crops via high capacity irrigation wells." While this is certainly troubling, Groves pointed out in a recent meeting of the Transdisciplinary Bee Deliberation, designed by Sai Suryanarayanan and Daniel Kleinman, that neonics were the only thing Huseth and Groves tested for. Our water table is probably filled with the residue of decades' worth of chemicals used in industrial farming, a veritable pesticide palimpsest. And if this is true in an agricultural region, what about the watershed surrounding suburban lawns?

The good news in all of this is that people can respond and make changes—and they have. Organizations like Friends of the Earth, the Pesticide Institute, and Bee Action have been encouraging the public to request neonicotinoid labeling on garden products, and some garden centers are providing neonic-free plants. The Xerces Society has been doing amazing work on many levels by encouraging homeowners to learn about their native plants and support their native pollinators. Xerces has been instrumental in planting drought-resistant flowers in almond groves, for example. More and more neighborhoods are choosing to be pesticide-free. Things can change if we make the change ourselves.

What would happen if we all stopped using gas-powered lawnmowers? What if we even went back to using push mowers? What if we cultivated an appreciation for dandelions and creeping Charlie? What if we tore up our acres of grass and planted pesticide-free wildflowers and vegetables? What if we stopped pouring chemicals and fertilizers on the grass? Think of the dent we could make in carbon emissions! Think of the healthy water in our lakes and watersheds! Think of the wonderful landscape for the bees! How do we give ourselves and our neighbors permission to stop these crazy and destructive practices?

The goal of my run that spring morning was to make it to the arboretum, which is lush with lilacs in the spring. I ran through the restored prairie, which, at the peak of summer, is a festival of color and abuzz with bees of all kinds. According to University of Wisconsin entomologist Claudio Gratton, there are more than five hundred species of wild bees living in Wisconsin alone. The UW Arboretum is a home to many of them. Gratton is a strong advocate of landscapes like this one, which support all kinds of insect biodiversity. We forget, I think, how important a rich tapestry of species is for a healthy ecosystem. Geoffrey G. E. Scudder, a zoologist from the University of British Columbia in Vancouver, put it this way: "Insects create the biological foundation for all terrestrial ecosystems. They cycle nutrients, pollinate plants, disperse seeds, maintain soil structure and fertility, control populations of other organisms, and provide a major food source for other taxa." Prairies like this one were important spaces.

In the spring, however, the lilac and magnolia trees draw all the attention, not just from bees but from people, too. There are always scores of winter-weary people wandering around the carefully mowed corridors between the trees, taking photos and pressing their faces into the blossoms to breathe deeply of their intoxicating scents. I never fail to bring my children to partake in this ritual of renewal, this awakening of life and return to the senses. One year, a good friend, the poet James Crews accompanied us on this journey, and afterward he wrote this poem, called "It Was Necessary":

> *Though we had sack lunches to pack*
> *and bags of dog food to haul home*
> *where dishes stood stacked in the sink,*
> *it was necessary for us to drive instead*
> *toward the arboretum, where lilacs*
> *at last in bloom filled the evening air*
> *with their perfume. It was necessary*
> *to get out of the car, kick off our shoes*
> *and walk barefoot onto the steaming,*
> *sun-warmed mulch among the bushes,*
> *to take a blossom-heavy branch*
> *in each hand and inhale deeply so we*
> *could breathe again after the longest*
> *winter in memory. It was necessary*
> *to close our eyes and sigh, to see*
> *what had been driving the bees so wild*
> *all day, making them dance above us*
> *as we collapsed in the dewy grass,*
> *half-high on that commonplace scent*
> *of spring we'd find hours later, clinging*
> *to our shirts, our skin, our hair.*

Time with the lilacs is indeed, in my opinion, absolutely necessary.

In the fall, the prairie next to the lilac grove I ran through all summer is cut and sprayed, as is necessary in prairie restoration projects, I understand. Signs dot the periphery of the fallen stalks and seeds of cup plants and coneflowers, the fallen branches and seeds of sumac. "Danger/Peligro: Keep Out. Garlon 4 Application," say the signs. Garlon 4 is a common herbicide that has not been proved to cause cancer, though the incidence of breast cancer increases when mice and rats are exposed to it. It causes kidney damage in dogs and adversely affects birds' eggs and frogs' behavior. For these reasons, perhaps, there is a warning sign to keep people out. But who will warn the animals? There are no blossoms for the bees to visit, but what of the others? Who will warn the birds and the deer and the worms?

William Eakin in collaboration with Aganetha Dyck, *Light*, 2011–12.

Trespass
The Art of Aganetha Dyck

Aganetha Dyck, a self-taught Canadian artist, was born in 1937 near Winnipeg, Manitoba. Dyck considers herself an environmental artist particularly interested in "the power of the small," and she has done many collaborations with honeybees. Much of her early work, shown all over Canada, England, France, and the Netherlands in the 1970s, was made with found objects like cigarettes and buttons, and was understood to be a critique of postwar domesticity. In 2000, she shifted her focus to bees and began to work with beekeepers and bees themselves to create honeycomb-covered objects. She is very interested in interspecies relationships.

The pieces included here are from a show called *MMasked Ball*, which the reviewer Dan Van Winkle called "horrifying" and "nightmare fuel." "The aim of Dyck's beehive art," says Van Winkle, "is to remind people just how intertwined our lives are with the nature that surrounds us, and we're all part of the same system. Mission accomplished. I will never again spend a waking minute where I don't feel like there are bees building me into some kind of horrible beehive monster." His review was accompanied by a short clip of a screaming man whose face was being swarmed by bees inside a cage. This unfortunate review—even if it was written partly in jest—perpetuates the fear that often surrounds bees and justifies the killing of them.

Fear of insects is nothing new. In an interview with Eric Brown, the editor of the brilliant essay collection *Insect Poetics*, Brown responds to a question about insects being seen as marvels or pests. "I would argue that the very things that annoy us about insects—their multiplicity, their voracity, their predation upon humans—are also most marvelous. . . . I'd say they remove us (sometimes harshly) from a quotidian existence and, like other innumerable—the stars, the sands on the beach—help us see the sublime in the everyday." The radical strangeness, the size, and the ubiquitous presence of insects make them perfect subjects for horror films, emphasizing the terrifying aspects of the sublime. Richard J. Leskosky argues that in the 1950s, oversized ants, grasshoppers, mosquitoes, and spiders allowed viewers to work through

Aganetha Dyck and the honeybees, from *MMasked Ball* (detail), series 2006–2008.

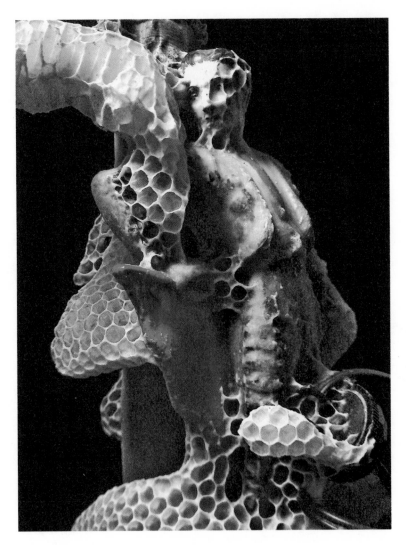

William Eakin in collaboration with Aganetha Dyck, *Light*, 2011–12.

Aganetha Dyck and the honeybees, from *MMasked Ball* (detail), series 2006–2008.

their anxieties about other pressing social issues, like the threat of nuclear attack. "In the case of big bug films," Leskosky writes, "viewers face not only their fears of creepy crawlers but also other more deeply-seated anxieties such as loss of life, stature and identity. The viewer similarly confronts in these movies anxieties over various other real-world phenomena such as threats posed by radiation, pollution, and the unintended consequence of scientific experimentation."

But what happens when, in real life, the insect becomes vulnerable, even endangered, as with honeybees and many other pollinators? It seems then that the perspective shift that occurs in the work of Aganetha Dyck actually becomes a way to empower the image of the fragile honeybee. To my mind, the collaborative bee-human project is a marvelous act of rebellion and trespass.

The genius of this particular group of pieces is that the bees have been given agency and have manipulated these symbols of high society from the height of the colonial enterprise. The porcelain tchotchkes of people dressed in eighteenth-century clothing enjoying a ball or having a tea party have been taken over and altered by bees. The pleasure that comes from this is a strange one, not unlike the pleasure one finds in reading the chapter of Alan Weisman's book *The World Without Us* in which he imagines New York City slowly reverting to wilderness after the human influence is gone. There is a kind of relief in believing that nature will regain control and heal itself, an avenue out of guilt, perhaps. Dyck's work seems to be both about honeybees adjusting to a human world, by forming their honeycomb on something as artificial as a porcelain sculpture of well-dressed women and men, and about the bees not respecting our rules, instead regaining power in spite of human endeavors to civilize and overtake nature, like obsessing about maintaining perfect lawns. Here, the honeybee is defiant, corrupting the pretty and destructive colonial aesthetic.

Another article about Dyck's work describes it as terrifying, as the bees appear to be entombing human bodies in wax. I would argue that Dyck does want the viewer to think of the bees as powerful, but also as necessary companions. Her research, she states, "asks questions about the ramifications all living beings would experience should honeybees disappear from earth." Their disappearance would be the most terrifying thing of all. Certainly, we would have no more opportunities to collaborate with them if they disappeared, either in the realm of art or in the realms of pollination and making honey. Dyck's bee sculptures tell a story of resilience, not unlike the stories I heard in Milwaukee, Wisconsin.

William Eakin in collaboration with Aganetha Dyck, *Light*, 2011–12.

Guard Bee

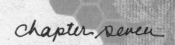

STORYING RESILIENCE

Get the fuck out of my garden! Get the FUCK *out of my* GARDEN! yelled a little kid hanging halfway out a school bus window as it slowly passed Alice's Garden, a community garden in a neighborhood in Milwaukee, Wisconsin. The garden filled almost an entire city block, and his voice floated easily over the chain-link fence, across the many neat rows of mint and beans, to reach our ears. Over and over, his commanding voice repeated the order as the bus traveled down one side of the block, turned, and then slowly crept down another.

Our host, Venice Williams, the visionary, charismatic director of the garden and its mission, stood in the center of a flowering labyrinth, consciously ignoring the boy's shouts, and continued telling us the story of the garden. Stories were of huge importance to Venice. In this way she reminded me of August, the sage beekeeper in *The Secret Life of Bees* by Sue Monk Kidd, who said, "Stories have to be told or they die, and when they die, we can't remember who we are or why we're here." As we stood in the middle of the garden, Venice told stories about a tradition of resistance, perseverance, and change. She recounted the story of Juneteenth Day, which was to be celebrated in just a few weeks—the day in 1865 when the news of the Emancipation Proclamation finally reached Galveston, Texas, and the enslaved were finally free. She told the story of how this once flourishing, almost entirely African American neighborhood had been torn apart when businesses and homes alike had been demolished in the name of progress to make space for the a new freeway, and stories of how people had been slowly rebuilding. She told stories about her journey to make this garden a place without segregation in a city that was famous for it, how she had forced Hmong garden-ers and African American gardeners to have plots side by side. She felt that the garden was healing the community. I felt that the storytelling itself was an act of resilience.

It was this neighborhood's vision of healing and resilience that had brought us, a primarily white, middle-class, interdisciplinary group from the University of Wisconsin–Madison, to Milwaukee. We were all part of a group called CHE, the Center for Culture, History, and the Environment of UW's Nelson Institute for Environmental Studies. Our plan was to investigate the workings of urban agriculture and to volunteer at both Alice's Garden and Walnut Way Conservation Corp., "a grassroots association committed to sustaining a diverse and neighborly community through civic engagement, economic development, and environmental stewardship." We wanted to spend a day or two helping turn over beds and dig up weeds, but we also wanted to learn why these projects had such a positive impact. Because I am a beekeeper, my own particular interest was in how bees were important to these organizations, both physically and metaphorically. The voice of the child on the bus reminded us of our status as visitors, even as we basked in the warm, welcoming voice of Venice.

After our brief introduction to Alice's Garden, we piled onto a bus to travel a few blocks to the house in Lindsay Heights where Walnut Way was headquartered. Stepping off the bus, I noticed the beehives right away. Faded blue and dull white hives, varying in size, clustered together in a side yard like the skyscrapers of a tiny city. A few bees hovered above the hives as others zoomed through the front entrances, loaded with pollen. The bees gave me a sense of comfort in this place where I was a guest.

Bees are hardly unique in urban settings. Beekeepers tend busy hives in New York City; San Francisco; Portland, Oregon; London; Seoul; and Cape Town. The rooftop of the National Assembly in Paris is home to three beehives. The Fairmont Hotel in Washington, D.C., has bees in its gardens who produce honey for its upscale restaurants. Bees float over city subways to find rooftop gardens and parks filled with black locust trees, lavender bushes, and squash plants. High-production urban farmers like Will Allen of Growing Power in Milwaukee need to keep bees to ensure that their vegetable plants are prolific. But bees at Walnut Way are more than honey producers or pollinators; they are also used as symbols.

The owner of the beehives was Larry Adams, a sturdy, middle-aged beekeeper with a bellowing laugh. He and his wife, Sharon Adams, have been the force behind the transformation of this neighborhood. Ten years or so ago, this street was filled with prostitutes and drug dealers, the houses were boarded up or falling down, and periodic gunshots were not uncommon. When Sharon tried to get insurance for her house, no one would insure her. Now there were yards filled with peach trees surrounded by rows of collard greens and tomato plants. Houses were getting new paint jobs, and crime was down. Sharon and Larry loved to share their success story. Inside a house that they had renovated and turned into Walnut Way's community center, the front room was filled with books and photos of gardeners. In a prominent spot on the wall hung a woodcut of a strong, African American man's face, which could easily have been Larry's, surrounded by honeycombs and bees. The words on the woodcut said "Bee Resourceful." Bees, I learned, are an emblem for this community. In one of her Walnut Way newsletters, Sharon writes, "Bees (and beekeepers) can teach us many lessons about stewardship. Their work ensures their community thrives. As stewards of our own environments, we all share the responsibility of making our communities wonderful places to live, learn, work, and serve." The bee is more than livestock for the Adamses. It is a symbol of the kind of stewardship that will make this neighborhood stronger and more resilient. The community cares for the bees, the bees pollinate the flowers, and the results are healthy vegetables and fruits, delicious honey, and healthier people. But this notion of stewardship also means taking care of this community of people, working hard, and working together, as bees do.

Walnut Way is certainly not the first group to choose bees as a model for a good society. For centuries, bees have appeared in narrative and myth as models of moral authority and ideal political systems. In the midst of Britain's Industrial Revolution, for example, George Cruikshank's illustration "The British Beehive" represented an acceptance of the social hierarchy of Victorian England. Cruikshank's image reflected natural scientists' perceptions of bees at the time—their view of each beehive as a tiny, industrious monarchy ruled by a queen. The idea of "industry" was both a moral imperative inspired by the Protestant work ethic and a powerful symbol of economic growth. The dynamics of labor were changing. The agricultural industry was changing. The number of farm laborers dropped by half between 1860 and 1900 as mechanization changed farming. Coal production was increasing. Large numbers of men and women were leaving the countryside and moving to urban centers to become employed in the cotton garment industry. This moment set the stage for the world we now occupy. The story of progress, equated with industrial improvements and faster, better systems, had begun.

Bees themselves were the subject of "progress" and "improvement." The question is whether the bee would be resilient enough to survive.

What exactly is a resilient community? There are many ways to answer this question. One way to think about resilience is to consider ecosystem health. For example, which species of fish can survive in a lake that is filling with phosphorous? I am particularly interested in resilience in an ecosystem that includes humans. Perhaps both the honeybees and the people in the neighborhoods of South Milwaukee are examples of resilient communities. How might their stories be integral to that resilience? Lisa Christensen and Naomi Krogman investigated how human communities survive under the pressures of resource extraction and climate change in the Yukon territory of Canada. Their research, based on the stories told in extensive interviews with community members, illuminates how the Champagne and Aishihik First Nations have been subjects as well as agents of change. While the people and bees of Lindsay Heights have not experienced the disturbing effects of an oil pipeline, climate change, or deforestation, they have suffered ruptures brought about by urban progress.

I recognize the dangers of comparing an insect community with a human community, but in this historical moment, the actions taken by humans and the existence or nonexistence of nonhuman communities are necessarily, and inextricably, entwined. This is especially true of a species like the honeybee, which has a rare symbiotic relationship with the human. And some of the same historical trends that have affected the honeybee have also affected the human. While human trends are changing, the stories that science is telling us about bees are also changing.

In the past, as in the illustration by Cruikshank, the honeybee was often seen as a model for a well-organized hierarchical society. This mirrored the assumption among scientists that bees were governed by the largest insect in the hive, the queen bee. But recently this notion has been overturned. The American scientist Thomas Seeley has been quietly studying bees' social organization for many years, and in 2011 he announced his findings. Through years of devoted observation of bees tagged with tiny numbers and colors, Seeley realized that the bees were making diplomatic decisions while they were swarming. A hive will swarm when the population has grown too large for its current home, and a small group of several hundred bees will leave with the queen to find new real estate. What Seeley observed was that the scout bees will return from different prospective new homes and dance a map to the others. The other bees will then go and check out the new location and return with their impressions. It is not until a majority of the bees decide in favor of one spot, which becomes evident in the fact that they are doing identical dances, that they make their move. That the queen was not making authoritarian decisions was groundbreaking news. The queen is obviously still essential, as she is the sole baby maker in the hive, but bees live in consensus communities. Seeley suggests that bees may have something to offer the human race in terms of fair and just decision making. Perhaps our new understanding of bees could lead to a new vision for human society.

Like the bees, Sharon and Larry Adams are interested in making changes that are born not solely in their own minds but in the dreams and needs of the neighborhood. They began by having potluck dinners, and, over pasta salads and pies, asking people to tell their stories. What did they remember about the neighborhood as children? What would they want to bring back? What did their grandmothers cook? What shops did they miss? What would they plant in their gardens? People used to sit on their porches in the evening, some people said. Others remembered peach trees. Everyone wanted an end to the violence. Many wanted a place where their children could go to read books and the parents could talk. Larry wanted to raise bees. To achieve these things, the neighbors had to agree to create boundaries for themselves. They had to participate in resistance against the forces that were keeping their community from becoming what it could be. They had to begin calling the police and reporting crime, refusing to let drug dealers onto their property, and standing up for the peaceful community they wanted to re-create. They had to attend the potlucks. And they were willing. They wanted change.

With these dreams in mind, Sharon and Larry established a relationship with Will Allen from Growing Power. In 1993, Allen, a former professional basketball player, had purchased an old greenhouse in northern Milwaukee, and forty acres that had been owned by his wife's parents, to create an enormous urban farm project. Allen transformed the old greenhouse into a vibrant space where nasturtiums curl down from elevated trays that are watered from tanks teeming with trout and tilapia beneath them, where mushroom chandeliers made from coffee bags hang from the ceiling, and where tanks filled with hundreds of red worms transform green waste into vermicompost. Behind the greenhouse, the quiet chatter of chickens and goats floats out over the piles of compost to the bee yard, which is vibrating with busy honeybees. This impressive, interconnected lattice of microindustries produces an abundance of healthy food, which is taken to local farmers' markets, served in Milwaukee's fine restaurants, and fed to people who live in the food deserts of the city. In addition to growing the food itself, Allen offers classes to children and adults on everything from aquaponics to food sovereignty. His project has drawn attention from people across the nation and has landed him many grants, including a leadership grant from the Ford Foundation and a genius grant from the MacArthur Foundation. He has become the Superman of the urban ag movement.

With Will Allen's inspiration, the residents of the Walnut Way neighborhood cleared trash from vacant lots and began building raised beds for vegetables, planting trees, and setting up beehives to pollinate the flowers. Artists painted murals on fences and walls. Now the neighborhood is flowering. As Sharon says, "People heal working collectively for the common good." Like Venice at Alice's Garden, Sharon beams as she tells the stories of resistance and improvement.

Her stories bring to mind another group of storytellers who take the bee as their symbol, a group called the Beehive Design Collective headquartered in Machias, Maine, but their stories are told with pictures. In August 2010, I sat outside a farmhouse in southwestern Wisconsin with artist Tyler Norman looking at the large, intricate mural printed on silk that he had spread out on the grass. The mural was covered with images of mammals, birds, and insects—all characters in an illustrated story of environmental degradation and human greed. Tyler is an illustrator and storyteller in a group of activist artists. The artists, who call themselves bees, hope to pollinate the world with ideas for radical changes in human behavior. They address such issues as pesticide use, mountaintop removal by coal companies, monoculture, and unfair labor practices. For them, the bee represents the ideal community member—going out to the fields, returning with news, and then heading out again. They travel from community to community, sharing their artworks and hoping to motivate people to change. Their intricate and finely detailed drawings depict animals of many kinds working in small collectives against huge machinery and faceless industrial giants that are extracting resources for money while destroying ecosystems and social networks. Salamanders work on bioremediation by planting cattails and oyster mushrooms to restore contaminated soil, honeybees harvest rainwater and plant gardens for community-supported agriculture, and turtles and raccoons with banners and signs spread the word about the grassroots organizations that are fighting to make the world cleaner and healthier and more just. As Tyler spoke, I recognized a tempered anger in his voice.

I thought of the boy on the bus, his voice like a fist in the air. He seemed to be voicing the fierce energy that is the stem of many movements, the energy that matures and is transformed into creativity and action. It occurred to me that I know very few children with that emphatic and proud sense of ownership in a plot of ground filled with vegetables and flowers. But this garden is special. It is a physical manifestation of a new moment in a movement against a long history of racism, of redlining, of destroying community.

Honeybees do not usually sting, because in the process of stinging they die, as the stinger is torn out of their abdomen and left in the skin of an intruder, but they do have guard bees, and eventually they will sting if they feel their community is in danger. Bears are well-known antagonists of bees. Bears break into a hive and destroy it in order to steal the sweet fruit of the bees' long labor. While the human relationship with bees is also one in which the bees are robbed of their honey, the mindful beekeeper takes only the surplus and works to preserve the health of the community. Guard bees warn a visitor to the hive, flying around the head of a careless beekeeper. I can attest to how a beekeeper's lack of mindfulness and respect leads to a bad interaction for all involved.

Like a guard bee, the boy on the school bus reminded me of the ways in which a community sometimes needs protection and mindfulness around boundaries. Boundaries of community and place are contingent upon the historical moment. We were standing in a community that had faced obliteration when bulldozers took out entire buildings to make space for a freeway, a community that had faced years of vicious racism, a community torn asunder by displacement and unemployment. This community was in the process of healing, and trust does not come instantly but must be earned.

The following afternoon, one of the residents of the Walnut Way neighborhood led us in some volunteer work. We talked as we bent over shovels and knelt pulling weeds in the garden skirting the meeting house. I guessed that this man was in his midfifties. He asked us many questions, perplexed by why we had come such a long distance by bus just to get dirt under our nails. When he learned that some of us were teachers and others students, he said, "Oh, so you're being graded for this. No wonder you're all working so hard." We laughed, but perhaps mostly out of embarrassment. No, we were here doing this labor because we craved a real connection to people and the earth. We were often stuck in rooms discussing ideas at length. We craved blisters on our hands. We craved sore shoulders. This felt good. It felt real. But we were not doing it in order to survive. We were labor tourists. As we worked, we told our own stories and became individuals. Some of us had our own gardens. I had bees. The man had been a construction worker but was happy to garden now. He showed us how to pull the crab grass with the full root system intact. He laughed at how slow I was, and eventually he relaxed, realizing that we really were there because we wanted to help.

As a beekeeper, I understand the need to protect a thriving but still vulnerable society. As my bees leave the hive each spring on an intrepid voyage to find nectar, they face a more and more perilous world. Under the auspices of production and development, there are fewer and fewer native flowering plants, as land is developed by industrial agriculture, pesticide use is ubiquitous, and GMO crops destroy global seed diversity. As the Beehive Design Collective illustrates in its murals, there are many forces at work that are not conducive to healthy living for bees or people. So the resistance of urban farmers and community gardens is not only of social significance; it is of ecological significance as well. These examples of small, diverse crops (not unlike those I saw in China), of low pesticide use, of beekeepers who understand the importance of bees and other pollinators to their food production, are examples of how we could move forward toward a more sustainable world.

When Christensen and Krogman went looking for resilience, they asked the Champagne and the Aishihik to tell them stories. In the stories being told at Walnut Way and Alice's Garden, we hear stories of change and resilience, but also of an imagined future.

On several Friday nights in the summer, Larry Adams and other folks who are interested in learning about bees light up the smoker to calm the bees, open up the hives, and pull out frames of honey. They uncap the comb and out pours the liquid light, the sweetness pulled from the neighborhood flowerbeds, peach trees, and cucumber blossoms, the sweetness that comes from so much hard work. Often, customers at the market ask Sharon, "Why is Walnut Way's honey so fragrant, sweet, and clear?" She responds, "This is what I know: the bees are loved. We provide safe hives, good water, and plenty of pesticide-free plants to sip nectar from. Larry always leaves enough honey-filled frames for the bees to eat and eat well." It is the story of Larry's honey, Sharon's peach trees, and Venice's neighborhood garden that I want to retell. The stories that are imagining and confirming the world I want to live in.

The Art of Resistance
The Beehive Design Collective

The "bee" I met from the Beehive Design Collective was only one of many artists working together to make the murals in this gallery. The collective has a core of ten to fifteen artists who work out of a building they own in Machias, Maine, but they very much want to steer clear of questions like "Who made this?" and "How much does this cost?" The work they produce is collaboratively made and copyright free. They are happy to share digital and print copies of the images for anyone who wants to pass them on. The murals develop around issues that require research. The first step in the process is gathering stories from people who are affected by the issues they want to address. They say of this process, "We gather stories for our graphics by first embarking on extensive listening projects, in order to more fully and accurately understand the complexity of the situation. From spending time in refugee camps in Latin America to visiting the coal fields of the central Appalachian mountains, we partner with communities and organizations who are seeking creative ways to amplify their voices and struggles for social and environmental justice." Once the stories have been gathered, the group members discuss ways to tell the story in images. Some of their pieces are very straightforward in their message, like the poster for a biojustice conference. Other pieces, like the mural called *The True Cost of Coal*, which uses animal and insect characters, tells the history of coal mining from the left side of the image to the right. The images in this gallery reflect the collective's position on biotechnology and plants that are bred with pesticides in them or require agrochemicals to survive.

The Beehive Design Collective is a good reminder of Margaret Mead's exhortation: "Never doubt that a small group of thoughtful, committed citizens can change the world. Indeed, it is the only thing that ever has." Grassroots organizations offer a viable path to shifting our story and creating the one we want. The biggest danger we face lies in not believing in our own power, in giving up. In a recent article, Rebecca Solnit addresses the problematic nature of presidential election years, which offer the public an opportunity to imagine that one person can change everything. One person can't. "Too many of us seem far too fond of narratives of our powerlessness," Solnit writes, "maybe because powerlessness lets us off the hook." Each of us can be consequential. We can change things. We just need to do it.

Two blueberry farmers, three urban gardeners, two entomologists, two wildlife biologists, and many beekeepers, artists, and poets . . . a good start. Let's add a mead maker who sees joy as a part of the solution.

Beehive Design Collective logo.

Beehive Design Collective, *The True Cost of Coal* (detail).

(opposite)
Beehive Design Collective,
Resist Biotechnology.

(left)
Beehive Design Collective,
Biodevastation.

A Different Kind of Buzz

chapter eight

MIRTH AS A FORM OF RESISTANCE

"What about joy?"

—JANE HIRSHFIELD

A cold wind and a sky the color of egg cartons greeted me outside the warehouse that is home to Bos Meadery the morning I met mead maker Colleen Bos for the first time. Inside the large, nondescript warehouse, I was surprised by hallways freshly painted white, hung with contemporary art. At the end of one long hall, from behind a steel door reading "Bos Meadery," came the sounds of '80s rock. At first, no one answered when I knocked, so I knocked again, harder. The music stopped, and the heavy door creaked open. Colleen's sister, Jeannine, a slight, dark-haired woman with bright eyes, in jeans and a T-shirt, peered out at me and invited me in. The room was not much bigger than a two-car garage, but one yellow wall and the warm wood floors made it feel spacious and bright. Several huge stainless steel tanks reminded me that this was indeed a facility that produced many, many bottles of mead. A counter on one side of the room was covered with empty glass bottles, bright as promises, beneath which clear plastic bags bulged with corks. Along another wall was what looked like a chemistry lab, with beakers and vials and tubes encased in a small glass house. I had so many questions.

Jeannine was preparing bottles for the bottling of their newest batch of mead, mead, I would learn, that was made from the honey of bees living in the city of Madison. A moment later, Colleen, the master mead maker, stepped through the door. She shook my hand and gave me a warm welcome. I had known enough not to expect my stereotype of a medieval mead maker—a short, round, jolly monk—but she had rosy cheeks, an infectious smile, and expressive hands, which did seem to fit my expectations somehow. I was already glad I'd come.

How exactly does one get into the mead business, I wondered. Colleen's interest in mead was sparked during her career as a medieval studies scholar. It was then that she first read about mead and mead halls in tales like *Beowulf*. She and her boyfriend, Peter, still enjoy sitting around and comparing translations of *Beowulf*, like those by Seamus Heaney and J. R. R. Tolkien. Though she had not intended to start a mead-making business, she was interesting in making the beverage and began home brewing. She and Peter shared their experiments with friends, and when a beekeeper suggested that she start a business making mead, she realized that this was exactly what she wanted to do. And indeed, it was exactly the right time and place. Colleen was now making mead for local restaurants and stores, and she had plans to expand.

Colleen walked me through the process of making mead. First of all, one needs honey. The honey Colleen uses in her mead comes primarily from beekeepers in Wisconsin. She has a strong desire for her mead to be a way to sample the flavors of a Wisconsin landscape. All of her beekeepers are "doing the right thing," she says, "like not taking the bees to the California almonds." The honey represents different seasons and different regions of Wisconsin. The most recent load of honey had come from the cranberry bogs up north, but the mead of the day was made with honey from a friend of hers at Mad Urban Bees, a concoction she was calling Mad Love.

The lab area is actually where her partner, Peter, spends most of his time. Here is where the nutrition of the yeast is measured and the yeast itself is propagated. They use a fermentation tank that works on the yeast the way sourdough starter does, with constant fermentation. Colleen imagines that the first mead was probably made by accident when some rain fell into a container of honey and fermented. Now she has much more control and is quite confident that her mead tastes much better than the beverage people were drinking in the Middle Ages. "This yeast is breathing our air and drinking our water. I love my yeast and feed them well," she said, eyes sparkling. This part of the process happens entirely under the "flow hood" to keep it consistent.

When you first mix the honey and yeast in the giant, 124-gallon fermenter, you only add the honey a little at a time over the course of a couple of weeks. In order to maintain maximum flavor and aroma, Colleen never heats the honey. This is a crucial point, because the varieties of mead depend entirely on the variations in honey. Wildflower honey makes a lighter mead, cranberry blossom honey is more tart, and buckwheat honey makes a mead that is more like Scotch whiskey, a very popular seller. At the end of the fermentation process, she might infuse the mead with hibiscus or ginger or some herb, depending on the batch.

The mead leaves the fermenter and enters the clarification tank for four to six weeks as the yeast flocculates out. Next, stabilizers are added, and then more honey. Then the mead is filtered, bottled, corked, and labeled, all by hand. The average batch makes forty-five cases of mead. From there it goes to local restaurants, bars, and stores.

Colleen says that her meads are all much drier than other meads on the market, which she believes gives her an edge. Because her mead is complex and flavorful, but not overly sweet, it pairs very well with specific foods. She loves the idea of the aromas and flavors of the Wisconsin countryside poured into a glass and paired with food grown in the same region. Each region, worldwide, has its own specific flavor.

Mead and mead halls appear in myths from all over the world, especially in medieval Nordic and Anglo-Saxon texts, and the drink is often the beverage of choice at Renaissance fairs today. One group of contemporary mead makers, Sky River Mead, claims on their website that "the very term 'honeymoon' comes from the ancient tradition of giving bridal couples a moon's worth of honey-wine. This was long ago thought to ensure a fruitful union." The Greeks believed that mead fostered good health and creativity. In *The Sacred Bee*, Hilda Ransome traces several moments of mead consumption and includes lines about Odin that make me feel hopeful. If only mead could do this for all of us:

> *A drink I took of the magic mead,*
> *Taken out of Othrörir.*
> *Then began I to know and to be wise,*
> *To grow and to weave poems.*

I smiled at this poem, but Colleen in fact believes that her mead has power. "It's a unique buzz," she said, smiling, like a person with a very special secret. "It's very joyous." She believes that the pleasure mead provides can help people fall in love with their local landscapes and encourage them to care more deeply about them. This is important, she said, because she, too, is worried about the bees.

Jeannine brought over two glasses containing a tiny bit of luminous gold liquid. "Sample?" she asked. We clinked our cups before raising them to our lips. I inhaled flowers and vanilla before the mead hit my tongue. Warm meadows filled my mind and a gentle heat radiated from inside my chest. In Colleen's joyful presence, I could almost believe that mead was magical, that it could make people fall in love with their landscapes and care for them more deeply. Inside that sunny room, the dark sky and the cold, and all the stories of doom outside seemed distant, even solvable. Was this joy a strategy for change, for survival?

Not long before my visit with Colleen, after a riveting lecture by Elizabeth Kolbert, the author of *The Sixth Extinction*, a brilliant, well-researched, and very bleak picture of the future of our planet, I asked Kolbert a question that may have sounded naïve. I said that I taught students who have been hearing the apocalyptic narrative since they were four years old, and that many of these young people seemed flattened out. I wondered what to tell them. Was there hope? Where did she find signs of it? She sighed and said she wished this hadn't been the first question, and that she basically felt that people needed to know the truth. As she spoke, I felt embarrassed for asking the question at all, and I felt she was right. In an interview with Sarah Dimick, Kolbert gave a very clear and reasonable explanation of her resistance to hope. She said, "If you listed what you would actually have to do to deal with these problems it would be laughable. Everyone would just say that is not happening. That was what I was faced with—this complete jam up—at the end. I needed a way out of it that was not on the one hand just throw up your hands in despair and on the other hand was not 'here, don't worry we can fix it.'" How can you possibly save the frogs from a global outbreak of chytrid, to name just one of the many challenges she lists? The truth is that when you're dealing with the sixth great extinction, there simply does not seem to be much hope.

I began to wonder if *hope* was the right word. Terry Tempest Williams had a great critique of the word *hope*, which was that it didn't have enough action in it. *Faith*, she felt, was a better word. In a conversation I had with the writer and feminist scholar Anne McClintock not long after Kolbert's lecture, McClintock said she preferred "questions of strategy." An artist from Scotland told me she preferred the word "possibility." The historian William Cronon said, on a panel the day after Kolbert's talk, "We cannot teach hopelessness." I told my father that I was struggling with language. "Hope," he said, "is necessary because it's the grease that keeps the wagon wheels turning." What I realized is that hope does seem necessary, but where and how could we cultivate it in a meaningful way? It didn't seem possible to work from a place of despair, so how could we instill joy, pleasure, or wonder in this bleak historical moment? How do we balance the need to tell devastating truths while still inviting people to find beauty and even mirth in life? Joy, pleasure, and wonder seem to be necessary components in our ability to connect to one another and to the world around us. And they are certainly important when trying to connect to something as small as a bee.

A couple of years ago, while I was a visiting lecturer in a section of environmental literature at the University of Wisconsin, the issue of hope arose during class. My plan was to get the students thinking about the relationships between humans and birds. During the first half of the class, we had discussed an essay on the "natural history" of the plastic pink flamingo by Jenny Price; for the second half of the class, I had hoped to get them more personally engaged with birds in their own lives. I asked them to explore the role that birds play, symbolically and actually, in their personal lives, in a ten-minute writing exercise. There were no restrictions on what they wrote. I suggested that they write whatever came to mind—whatever memories, metaphors, irritations, dreams, scientific facts, or stories about birds occurred to them. I prepared them for this exercise with three short experiences. The first was reading aloud Susan Stewart's poem "Wings," in which one voice insistently asks another, "If you could have wings would you want them," even if they're large and heavy and you couldn't take them off? "Yes," the other voice answers, just as insistently, "I would still want them . . . because of the flying." Next, I showed them a series of extraordinary photos of bird nests by Sharon Beals. Then I had them watch a three-minute YouTube clip of birds in flight with no narration. They wrote busily for ten minutes, and then I asked them to choose one sentence or phrase from their piece to read aloud, one by one, until everyone had contributed a line or thought. In the process, we created a poem, "23 Ways of Looking at Birds." It was a surprising, diverse, and wonderful poem, and a great snapshot of how birds exist in myriad ways in our consciousness.

But then, because I wanted to show them how awful birds have it these days thanks to reckless human consumption, I showed them a trailer for Chris Jordan's 2011 film *Midway*, which offers an intimate portrait of an island near the Great Pacific Garbage Patch inhabited by albatross who are suffering and dying because of the great quantity of plastic they ingest. Several shots displayed decaying bird carcasses, their eroded stomachs bulging with tiny plastic objects. One shot captured a baby albatross writhing in her last spasms of pain before dying. Afterward, I asked them to respond. They were completely stunned and seemed almost despondent, speaking only as if they felt obligated to respond. This surprised me, because I had assumed they might react with outrage and an impulse toward activism, or perhaps I thought some would be jaded and unmovable. Instead, perhaps because I had brought them to such a vulnerable place beforehand, they were devastated. I asked them if they found the trailer successful, and whether they wanted to see the film, or to share it with others. A few said they would; many said they wouldn't. One person said it made her feel hopeless. A male student seemed so deflated, in fact, that I suddenly felt horrible for showing the clip at all. He managed to say, "Birds always represented freedom to me, and they aren't free. They're going to be wiped out by the humans, too."

I realized that this might be a great opening for a discussion about how stories can motivate or extinguish activism. About this, I was relieved and grateful to learn, they had a lot to say. There were many opinions about reaching different audiences with different methods—ideas about offering solutions rather than only illustrating problems, but the recognition that both were necessary. Some liked the specificity of the cigarette lighters and bottle caps in the bird bellies because they felt personally implicated. Many felt overwhelmed, seeing the Pacific Garbage Patch as a disaster that seems too big to fix. I was fascinated, and I started to wonder whether these young people were growing up in a time when apocalyptic narratives were not simply scare tactics but simply reality. Some generations, it seems, need to be shocked into awareness. It occurred to me that maybe this generation did need hope. Not an empty, vacuous kind of hope, but hope built on the possibility of actual change. We talked then about how that kind of hope is or is not built. I told them that an environmental scholar named Nancy Langston had told me that she once taught an environmental studies class that some of her students called, "We're Fucked 101," and how she had realized that it was important for students to know that they are not powerless but can actually do something to make things better. But how do you tell the depressing truth and also inspire action and innovation? At the end of our conversation, the students left the room quietly.

The next time I met with them, the subject of the class was bees. I had a different plan this time, and I began by asking them to write down all of the words that came to mind when I said *bugs* or *insects*, and then had them read their lists aloud. I imagined that many folks wouldn't have a lot of good things to say; many, as I suspected, were afraid of spiders and hated mosquitos but loved butterflies. We talked briefly about the idea of insects as pests and about how their size makes it easy to kill them. Then I began talking about bees and the long history of human relationships and dependence on them, about how we have revered bees as gods, about how we have used them as political symbols, and about how we have been using them for pollination at least since ancient Egypt. This led to a discussion of bees' remarkable senses, and finally to the dangers they face from our current pesticide use and some of the ways in which people are trying to change these practices. At this point, to immerse the students more deeply in the world of bees, I passed around paper bags full of things like fresh lavender, coffee, and mint leaves and had them smell each of the different scents. Imagine experiencing your geography the way a bee must, I suggested, by navigating through one intense smell after another. Next, I passed out several bouquets of flowers and handed out magnifying glasses. Look into these flowers and imagine flying in and out of them, I suggested. As they were staring and observing, I passed around samples of cranberry honey, buckwheat honey, and clover honey for them to taste. At the end of the class, I had them write a short, anonymous reflection. Here are a few of the things they wrote:

> "I never knew honey had different flavors based on the flower. The best one, I thought, was white clover—far and away. It was euphoric really."

> "I really enjoyed smelling the lavender. It was pleasant, and I realized that I don't smell things enough."

> "This is the first time in awhile that I've actually looked into a flower. I'm used to simply passing by them. . . . It's a good reminder that the world doesn't always need to be so big. Sometimes taking a step back and just experiencing the place you are in can do a lot. The bee perspective might be the right one."

> "The movement to reduce pesticides . . . is crucial."

> "Bees are truly a keystone species."

Two people even said they would consider keeping bees in the future! My tactics were not entirely different for the two classes, but one definitely felt more successful than the other. I wanted students to think more deeply about their personal relationships with nonhuman others, to create an environment that would cultivate intimacy and mindfulness, and then to reflect on the realities for those others in this historical moment. But by delivering the horrific news at the end of the first class and offering no strategies for change, I had invited despair. Luckily, this was a group of students who would talk with me about navigating through the dark cloud I had introduced. After the bee class, however, even after learning about the problems bees are facing, students left with some buoyancy, even joy.

I have no way of knowing whether a joyful attitude actually makes a person do more good in the world. I know that depression makes progress of any kind difficult for most people, and impossible for many. A recent psychological study of climate scientists has shown that they experience something akin to post-traumatic stress disorder; some are calling it "pre-traumatic stress disorder." Knowledge of what lies ahead keeps them from being able to stay hopeful in the present. As Lise Van Susteren, the psychiatrist who coined the term, puts it, "when you're talking about thousands of years of impacts and species, giving a shit about whether you're going to get the right soccer equipment or whether you forgot something at school is pretty tough." One climate scientist said she wouldn't return to certain sites because she didn't want to face how much worse they'd become.

I don't think drinking mead will necessarily create a love of landscape. But I do know that in the presence of someone like Colleen Bos, who was teaching me how to understand the wonder of mead and the landscapes that made it possible, I did feel a sense of—I'll say it—hope. And it brought to mind Wisconsin's 2013–14 poet laureate Max Garland's words:

> Here in the diminished light of harvest,
> though Holsteins graze the hillsides
> and cranberry bogs are bursting red,
> it's by cultivating wonder
> the commonwealth is fed.

Bee Renaissance
The Art of Lea Bradovich

Lea Bradovich is a painter living in Santa Fe, New Mexico, whose work has been shown in many galleries in the United States. She says that she uses myth and natural imagery in her work. Inspired by Renaissance painters, her beautiful portraits share tone and compositional elements with painters like Jan van Eyck and Raphael. The paintings in this gallery are portraits of humans and bees.

What drew me to these images most forcefully was their complex ambiguity. One might read these paintings as images of people using bees as decoration, which certainly seems to reflect our current use of the bee as a tool or a symbol divorced from its ontology. But another way to see them— indeed, the way I think Bradovich intends us to see them—is as images of humans honoring the bee, perhaps representing something like a bee renaissance. The bees are quietly working the comb in several of the images, the human and the bee both seemingly at peace. What I choose to see is coexistence.

A movement toward coexistence is what I have encountered over and over

on this journey of learning about bees and the people who love them. Nonetheless, I recently read an article confirming another 37 million bees dead in Canada. The article was careful to avoid drawing definite conclusions as to what killed them, but the beekeeper quoted in the piece is convinced that neonicotinoids are the culprit. Maybe the bees won't win this fight. Maybe the electronic bee will be a godsend. But many artists, scientists, beekeepers, farmers, activists, and citizens are creatively advocating for the bees; a few of them are featured in this book. In a healthy beehive, labor is divided among the bees, some gathering nectar and pollen, some nursing new bees, and some building comb as others produce honey and wax. To change the way we live on this planet will take everyone doing what they are best at doing. Writing this book has provided scaffolding for my belief in the human potential for creativity, connection, compassion, wonder, cooperation, joy, and action. Is it possible that we are on the verge of an environmental renaissance? Time will tell.

Lea Bradovich, *Daughter of the Hive.* Acrylic
gouache on panel, 24 × 18 in.

Lea Bradovich, *Queen Bee*. Acrylic gouache
on panel, 12 × 9 in.

Lea Bradovich, *Look Homeward Queen Bee.*
Oil on panel, 18 × 14 in.

Lea Bradovich, *Queen's Regalia.*
Acrylic gouache on panel, 24 × 18 in.

Afterword

Tuesday, 7:30 a.m. A mug of honey-sweetened coffee warms my hands. A cool wind is moving big cumulus clouds across the sky this morning. The patches of light passing through the maples brighten and fade as the sun appears and disappears. The honey jars on my windowsill are illuminated briefly with each brightening of the sky. The air reminds me that another summer is ending. I step outside and stroll over to one of my hives. A handful of bees snuggle near the opening. They have produced so much honey this year that I was able to take a bit for myself. I feel good about their going into the winter with so much to nourish them. Midwestern winters are harsh.

I remember one strangely warm winter day, when the temperature in Wisconsin had climbed to eighteen degrees after days of wind-chill readings of fifty below. I walked out to the hives, assuming that the bees were dead. The bees in the first hive had looked weak in October. Though frames of their hard-won gold glistened in the light when I pulled them, their numbers had seemed too low. A swarm had left these bees behind to rebuild their tiny city. Then some of the neighbors sprayed their yards with pesticides. In January came a wind like no one could remember. Cows' ears cracked like glass in the cold. Nothing escaped that wind. The day it warmed to eighteen above zero, I laid my head against the hive. No sound, as I suspected. But I knocked anyway, hard, several times, against the side of the box, and then I heard the roar, the gathering roar of the huddled survivors alarmed by my assault, the quiet roar of fire. To be honest, when I heard them, I wept, moved by their incredible resilience. It was a very good reminder that together, and with love, we can survive just about anything.

But today, winter is still far off. I let a honeybee crawl up on my hand, her little feet gripping my skin. I look into her deep dark eyes. Her antennae explore the air, my hand. This little bee, who is capable of pollinating crops, making honey, building cathedrals of wax, and who is living in a society at risk, is completely composed on my finger. For a moment, she rearranges her diaphanous wings, and then she lifts into the sunlight, carrying on.

Notes

The notes below are keyed to the text by page numbers.

Introduction

2   *The Berkeley geographer*: See Kosek, "Ecologies of Empire."

At Inscentinel: Panchromos Limited, "Inscentinel Vasor."

A designer named Susana Soares: Susana Soares, "Bees," accessed October 8, 2016, http://www.susanasoares.com/index.php?id=52.

The bees' pollen was tested: Sewell, "Gold Miners All A-Buzz."

3   *"It takes a cell full of honey"*: Woller, interview by author.

"The 'potentially disastrous' decline": UN News Centre, "Humans Must Change Behaviour."

Bees have been moved: Crane, *World History of Beekeeping*, 472.

4   *A USDA study*: LaJeunesse, "Diet Affects Pesticide Resistance."

The bees and many other pollinators…: Nixon, "Slow Violence and Environmental Storytelling."

5   *"The U.S. scientific community…"*: See http://www.merchantsofdoubt.org/.

Rudolf Steiner, among others, argues: See Steiner, *Bees*.

6   *"In the U.S. alone…"*: Schiffman, "Mystery of the Missing Bees."

The "New York Times" reported: Chang, "Stanford Scientists Cast Doubt."

8   *For him, the bees are spiritual teachers*: Thiele, interview by author.

Chapter 1

13   *"I will arise and go now"*: William Butler Yeats, "The Lake Isle of Innisfree," in Untermeyer, *Modern British Poetry*, 120.

14   *According to the National Institutes of Health*: Golden, "Insect Sting Anaphylaxis."

16   *"It seems natural…"*: Dominus, "Mystery of the Red Bees."

17   *"Dear Sir…"*: Dixie Larkin, on behalf of the Committee of a Thousand, to Wisconsin Conservation Department, July 1957, Wisconsin Historical Society, Wisconsin Conservation Department Files, 1957, http://www.wisconsinhistory.org.

18   *"We implore you…"*: Marie Thompson, president of the Animal Protective League, to John Beale, chief forester, Wisconsin Conservation Department, June 30, 1957, ibid.

"From the beginning…": Nelson and Surber, "DDT Investigations."

"Carbaryl [or Sevin] is one…": U.S. EPA, "Carbaryl IRED Facts," 1.

20   *"We abuse land…"*: Leopold, *Sand County Almanac*, xviii–xix.

27   *"I explore the lack of harmony…"*: Sibylle Peretti, "Artist Statement,"

http://sibylleperetti.com/artwork
/3800291-Artist-Statement.html.

27 She said later that she "felt obligated ...":
Quoted in Dhruba, "Uncovering Interior
Dialogues."

28 "A child's world is fresh ...": Carson,
Sense of Wonder, 22.

Chapter 2

32 "The birds have vanished ...": Hamill,
"Zazen on Ching-t'ing Mountain."

33 I wondered whether the Grain for Green
program: See Chen et al., "Factors
Affecting Land Reconversion."

35 Their solution was pollinating by hand:
Partap and Ya, "Human Pollinators of
Fruit Crops."

36 "Blossoming, discreetly ...": Murong,
"Blossom Tree."

38 "Can the [Chinese] government ...":
Schiavenza, "Dead Swine-Gate."

That's a return to the old ways ...: BBC
News, "Billion Go Hungry."

41 "The passion caused ...": Burke, Sublime
and the Beautiful, 79.

"[T]he Alps, abysses ...": Cuddon,
Dictionary of Literary Terms, 875–76.

"The dream of flight ...": Holmes, Age of
Wonder, 125.

44 "[O]ver time, we will only get better ...":
Bailey, "Better to Be Potent Than Not."

"But for others ...": Nixon,
"Anthropocene."

47 Lacking an understanding of depth:
Dillard, "Seeing," 27–29.

"Whereas the miniature represents closure": Stewart, On Longing, 70–71.

Chapter 3

49 "The future of all wildlife ...": Moritz
Zimmermann, Save the Elephants:
Securing a Future (2010), https://
miguelcuespinosa.wordpress
.com/2013/10/21/save-the-ele
phants-by-moritz-zimmermann/.

50 In 2002, Iain Douglas-Hamilton and Fritz
Vollrath ...: See Douglas-Hamilton
and Vollrath, "African Bees to Control
African Elephants."

51 San Honey Hunter's Prayer: "Hunter's
Prayer," quoted in Crane, World History
of Beekeeping, 53.

"Particular magico-religious importance ...":
Pager, Stone Age Myth and Magic, 74.

52 A San person explained ...: Quoted in
Lewis-Williams, "Thin Red Line," 8.

53 "The cocktail of pesticides ...": See
Penn State Extension, "Pesticides and
Pollinators."

"Some agricultural pesticides": Hubbell,
Book of Bees, 136.

54 "It is not an exaggeration ...": Quoted in
Begg, "Bees."

55 "One of the most significant problems":
Hitchcock, Kalahari Communities, 10.

In looking specifically at the situation of
the San: Ibid., 13.

The ARC website reports: Agricultural
Research Council, "Beekeeping for
Poverty Relief Programme."

One tourism website: "Experience Africa
with Rob and Stella, South African

citizens now living in the USA. We invite you to join us on our unique tours and wildlife safaris to Southern Africa. We speak the local languages and have extensive knowledge of the African Culture and the country as a whole. We are both certified South Africa Fund Tourism Experts certified by South African Tourism. Enjoy excellent cuisine together with superb South African wines. Great souvenir shopping, African art, skins and hides, diamonds, and Kruger Rands." Boomerater, "International Beekeeping Safari 2009," accessed October 24, 2016, http://www.boomerater.com/travel /vacation-package-830-internation al-beekeeping-safari-2009-trip-idea.

56 "In this actual world ...": Williams, "Ideas of Nature," 83–84.

57 Globalization invites universalizing solutions: See Tsing's book Friction.

60 "My core idea ...": Kim Gurney to author, December 15, 2010.

66 "Dismembered beehive frames ...": Kim Gurney, Frugi Bonai, exhibition catalogue from Gurney's exhibit in the Artspace Gallery, Johannesburg, South Africa, November 11–December 2, 2009, 11.

Chapter 4

68 "The god Indra ...": Gough, "Bee: Part 2—Beewildered."

He began by telling me ... : Suryanarayanan, interview by author. Additional statements by Suryanarayanan in this chapter are from this interview.

"'Anthropomorphism' is the word": Daston and Mitman, "Introduction," in Thinking with Animals, 2.

70 "Watching him, it seemed as if a fibre ...": Woolf, "The Death of the Moth," in Death of the Moth, 4, 5.

As many of us know, the waggle dance: See von Frisch, Bees.

71 "Using high-speed video techniques ...": Gil and De Marco, "Decoding Information," 887.

"It's bad biology ...": Bekoff, Animals Matter, xviii.

72 The experiment they designed ... : Bateson et al., "Agitated Honeybees," 1070.

73 "Whoever ... should be guilty ...": Quoted in Asdal, "Subjected to Parliament," 901–2.

74 "[E]thologists who have devoted their lives ...": Daston and Mitman, "Introduction," in Thinking with Animals, 7.

75 "They were so cute ...": Stack-Whitney, interview by author. Stack-Whitney's statements in the next five paragraphs are from this interview.

78 "[D]ominant approaches ...": Kleinman and Suryanarayanan, "Dying Bees," 494–95.

"Despite the conclusion ...": Kleinman and Suryanarayanan, "Honey Bees Under Threat."

79 "Empathy, wonder ...": Whitney, "Tangled Up in Knots," 106.

81 "I am trying to resist ...": Quoted in Katcher, "Art and Animals."

82 "[T]he return of taxidermy ...": Aloi, "Rogue Taxidermy," 2.

82 *Morgan's art, by contrast*: See Aloi and Frank, "In Conversation with Claire Morgan."

"We do not need or desire souvenirs …": Stewart, *On Longing,* 135.

84 *The environment in which they are glued*: See the works on Hatton's website, http://sarahhattonartist.com/bees/.

Chapter 5

86 *"To make a prairie …"*: Emily Dickinson, "To Make a Prairie (1755)," https://www.poets.org/poetsorg/poem/make-prairie-1755.

87 *"Now the number of mice …"*: Darwin, *Darwin,* 74.

"Our results suggest": Brosi and Briggs, "Single Pollinator Species Losses," 13044.

"In the space of this garden …": Pollan, "Border Whores."

89 *"Scale is one issue …"*: John and Christina Eisbach, interview by author.

90 *Michigan State University professor of entomology …*: Rufus Isaacs, interview by author.

91 *"We love bees"*: Dennis Hartmann, interview by author.

93 *"As you have already been told …"*: Parker and Gregg, *Insect Friends and Enemies,* 36.

94 *"They estimated that at any given time"*: Raffles, *Insectopedia,* 7.

In a fifteen-minute documentary: "DDT—Weapon Against Disease" can be viewed on YouTube at https://www.youtube.com/watch?v=IJCdXNl2Duc.

97 *"The decorative details"*: For more on Goluch's work, visit her website at http://www.elizabethgoluch.com.

The tiny robotic bees …: See Harvard's John A. Paulson School of Engineering and Applied Science (SEAS) website at http://www.seas.harvard.edu. See also Perry, "Robotic Insects."

99 *"One day long after we are dead …"*: Sheck, "One Day," 77.

Chapter 6

104 *Their most modest estimate …*: Milesi et al., "Mapping and Modeling the Biogeochemical Cycling."

"Even conservatively": Quoted in Earth Observatory, "More Lawns Than Irrigated Corn."

The USDA produced a report …: USDA, National Agricultural Statistics Service, *Vegetables: 2015 Summary.*

"Covering a total area …": Robbins, *Lawn People,* viii.

Milesi's team estimated …: Polycarpou, "Problem of Lawns."

A 2008 article …: *Scientific American,* "How to Pick a Lawn Mower."

105 *The National Gardening Association boasted*: National Gardening Association, "2013 National Gardening Survey."

"Lawn pesticides are applied": Robbins, *Lawn People,* xiii.

In 2000, the U.S. Fish and Wildlife Service claimed: U.S. Fish and Wildlife Service, "Homeowner's Guide to Protecting Frogs."

105 *Dicamba toxicity in animals ...*: National Pesticide Information Center, "Pesticide Fact Sheets."

106 *They quote an EPA estimate*: Bormann, Balmori, and Geballe, *Redesigning the American Lawn,* 74.

The state's definition of agricultural use: California Department of Pesticide Regulation, "Pesticide Use Reporting."

The colonists brought with them ...: Bormann, Balmori, and Geballe, *Redesigning the American Lawn,* 18.

107 *"The bees have generally extended themselves ..."*: Quoted in Horn, *Bees in America,* 5, 41.

108 *"Such a lawn only developed ..."*: Robbins, *Lawn People,* 129–30.

A study released in June 2014: Friends of the Earth, "Gardeners Beware 2014."

109 *In a survey of one agricultural region ...*: Huseth and Groves, "Environmental Fate of Soil."

According to University of Wisconsin entomologist Claudio Gratton: See Gratton Lab: Landscape Ecology of Insects and Arthropods, University of Wisconsin–Madison, http://gratton .entomology.wisc.edu/pollinators/.

110 *"Insects create the biological foundation ..."*: Scudder, "Importance of Insects," 9.

113 *Aganetha Dyck, a self-taught Canadian artist ...*: For more on Dyck's work, visit her website at http://www.aganetha-dyck.ca.

The pieces included here ...: Van Winkle, "Artist Uses Real Bees."

"I would argue that the very things that annoy us ...": Aloi, "Talking Insects—Eric Brown," 6.

117 *"In the case of big bug films ..."*: Leskosky, "Size Matters," 319–20.

Chapter 7

119 *"Stories have to be told"*: Kidd, *Secret Life of Bees,* 107.

120 *"[A] grassroots association"*: Walnut Way Conservation Corp. website, accessed 15 September 2013, http://www.walnutway.org.

High-production urban farmers: See "Growing Power: About Us," accessed November 2, 2016, http://www.growingpower.org.

121 *"Bees (and beekeepers) can teach us ..."*: Walnut Way Conservation Corp., http://www.walnutway.org. All quotations from Sharon Adams are from this website.

In the midst of Britain's Industrial Revolution ...: See Atterbury, "Steam and Speed."

122 *Their research, based on the stories told*: See Christensen and Krogman, "Social Thresholds and Their Translation."

Through years of devoted observation ...: Seeley, *Honeybee Democracy,* 218–31.

124 *Her stories bring to mind ...*: For more on the Beehive Design Collective, visit http://beehivecollective.org.

In August 2010, I sat outside a farmhouse ...: Norman, interview by author.

127 *"We gather stories ..."*: Beehive Design Collective, "What We Do," http://

beehivecollective.org/about-the-hive
/what-we-do/.

127 *"Never doubt that a small group"*:
Margaret Mead's statement can be
found at http://womenshistory.about
.com/cs/quotes/a/qu_margaretmead
.htm.

128 *"Too many of us seem far too fond"*:
Solnit, "One Magical Politician."

Chapter 8

133 *"What about joy?"*: The epigraph to this
chapter is from Hirshfield, "Salt Heart,"
14.

134 *All of her beekeepers …* : Colleen Bos,
interview by author. All quotations
from Bos are from this interview.

135 *One group of contemporary mead makers
…* : Sky River Meadery, "A Brief History
of Mead," accessed October 23, 2016,
http://www.skyriverbrewing.com/
Mead/mead-history.html.

"A drink I took …": Ransome, *Sacred Bee*,
158.

136 *"If you listed …"*: Dimick, "Speaking to
Us."

*Terry Tempest Williams had a great
critique …* : Terry Tempest Williams,
Jordahl Public Lands Lecture, University

of Wisconsin–Madison, October 22,
2013.

"We cannot teach hopelessness": William
Cronon, Anthropocene Slam: A
Cabinet of Curiosities, Deluca Forum,
Wisconsin Institutes of Discovery,
Madison, November 10, 2014.

"'Hope,' he said, 'is necessary'": Arthur
Kdav, interview by author.

137 *Next, I showed them a series*: Beals's
photos can be viewed at http://
www.sharonbeals.com/fine-art/#/
nests-part-one/.

But then, because I wanted to show them:
For more on Jordan's film *Midway:
Message from the Gyre*, including the
trailer, see http://www.midwayfilm
.com/ or https://vimeo.com/25563376.

139 *A recent psychological study …*: Quoted
in Thomas, "Climate Depression Is for
Real."

140 *"[H]ere in the diminished light of harvest"*: Garland, "For a Dedication by the
River."

141 *The article was careful*: Ballingall,
"Ontario Honey Bees Dropping Like
Flies."

Bibliography

Published Sources

Agricultural Research Council. "Beekeeping for Poverty Relief Programme." Accessed October 24, 2016. http://www.arc.agric.za/arc-ppri/Pages/Insect%20Ecology/Poverty-Relief.aspx.

Ahuja, Neel. "Postcolonial Critique in a Multispecies World." *PMLA* 124, no. 2 (2009): 556–63.

Aloi, Giovanni. "Rogue Taxidermy: Editorial." *Antennae: The Journal of Nature in Visual Culture* 6 (Summer 2008): 2.

———. "Talking Insects—Eric Brown." *Antennae: The Journal of Nature in Visual Culture* 1 (Autumn 2007): 4–7.

Aloi, Giovanni, and Eric Frank. "In Conversation with Claire Morgan: Interview by Giovanni Aloi and Eric Frank." *Antennae: The Journal of Nature in Visual Culture* 6 (Summer 2008): 55–59.

Asdal, Kristin. "Subjected to Parliament: The Laboratory of Experimental Medicine and the Animal Body." *Social Studies of Science* 38, no. 6 (2008): 899–917.

Atterbury, Paul. "Steam and Speed: Industry, Power, and Social Change in Nineteenth-Century Britain." Accessed September 30, 2016. http://www.vam.ac.uk/content/articles/s/industry-power-and-social-change/.

Bailey, Ronald. "Better to Be Potent Than Not." *New York Times,* May 23, 2011. http://www.nytimes.com/roomfordebate/2011/05/19/the-age-of-anthropocene-should-we-worry/better-to-be-potent-than-not.

Ballingall, Alex. "Ontario Honey Bees Dropping Like Flies: Beekeeper Blames Seed Pesticides." *Toronto Star*, July 5, 2013. https://www.thestar.com/news/gta/2013/07/05/ontario_bees_dropping_like_flies_elmwood_beekeeper_blames_corn_pesticide.html.

Bateson, Melissa, Suzanne Desire, Sarah E. Gartside, and Geraldine A. Wright. "Agitated Honeybees Exhibit Pessimistic Cognitive Biases." *Current Biology* 21, no. 12 (2011): 1070–73. https://www.ncbi.nlm.nih.gov/pmc/articles/PMC3158593/.

BBC News. "A Billion Go Hungry Because of GMO Farming: Vandana Shiva." April 4, 2013. https://www.youtube.com/watch?v=vbIQF72IDuw.

Begg, Angus. "Bees." *Carte Blanche*, MNET, May 30, 2010. http://beta.mnet.co.za/carteblanche/Article.aspx?Id=3975.

Bekoff, Marc. *Animals Matter: A Biologist Explains Why We Should Treat Animals with Compassion and Respect.* Boston: Shambhala, 2007.

Bishop, Holley. *Robbing the Bees: A Biography of Honey—The Sweet Liquid That*

Seduced the World. New York: Free
Press, 2005.

Bormann, F. Herbert, Diana Balmori,
and Gordon T. Geballe. *Redesigning
the American Lawn: A Search for
Environmental Harmony.* 2nd ed. New
Haven: Yale University Press, 2001.

Brosi, Berry J., and Heather M. Briggs.
"Single Pollinator Species Losses
Reduce Floral Fidelity and Plant
Reproductive Function." *Proceedings of
the National Academy of Sciences* 110, no.
32 (2013): 13044–48.

Brown, Eric C. "Introduction." In *Insect
Poetics,* edited by Eric C. Brown, ix–xxiii.
Minneapolis: University of Minnesota
Press, 2006.

Burke, Edmund. *A Philosophical Inquiry into
the Origin of Our Ideas of the Sublime
and the Beautiful.* New ed. Basel: Printed
and sold by J. J. Tourneisen, 1792.

Burns, Deborah, ed. *Attracting Native
Pollinators.* North Adams, Mass.: Storey,
2011.

Burns, Loree Griffin. *The Hive Detectives:
Chronicle of a Honey Bee Catastrophe.*
New York: Houghton Mifflin, 2010.

Burtynsky, Edward, Lori Pauli, Mark
Haworth-Booth, Kenneth Baker, and
Michael Torosian. *Manufactured
Landscapes: The Photographs of Edward
Burtynsky.* New Haven: Yale University
Press, 2003.

California Department of Pesticide
Regulation. "Pesticide Use Reporting."
Accessed October 24, 2016. http://
www.cdpr.ca.gov/docs/pur/purmain
.htm.

Carson, Rachel. *The Sense of Wonder.* New
York: Harper Collins, 1965.

———. *Silent Spring.* New York: Houghton
Mifflin, 1962.

Chang, Kenneth. "Stanford Scientists Cast
Doubt on Advantages of Organic
Meat and Produce." *New York Times,*
September 3, 2012.

Chen, Xiaodong, Frank Lupi, Guangming
He, Zhiyun Ouyang, and Jianguo Liu.
"Factors Affecting Land Reconversion
Plans Following a Payment for
Ecosystem Service Program." *Biological
Conservation* 142, no. 8 (2009): 1740–47.

Christensen, Lisa, and Naomi Krogman.
"Social Thresholds and Their
Translation into Social-Ecological
Management Practices." *Ecology and
Society* 17, no. 1 (2012). http://www
.ecologyandsociety.org/vol17/iss1/art5/.

Crane, Eva. *The World History of Beekeeping
and Honey Hunting.* New York:
Routledge, 1999.

Cuddon, J. A., ed. *A Dictionary of Literary
Terms and Literary Theory.* 4th ed.
Oxford: Basil Blackwell, 1991.

Darwin, Charles. *Darwin.* Edited by Philip
Appleman. New York: W. W. Norton,
1970.

Daston, Lorraine, and Gregg Mitman, eds.
*Thinking with Animals: New Perspectives
on Anthropomorphism.* New York:
Columbia University Press, 2006.

Dhruba, R. C. "Uncovering Interior
Dialogues: An Interview with Sibylle
Peretti." *Lucky Compiler,* May 15,
2013. http://luckycompiler.com
/uncovering-interior-dialogues/.

Dillard, Annie. "Seeing." In *Pilgrim at Tinker Creek*, 16–36. New York: Harper Perennial, 1974.

Dimick, Sarah. "Speaking to Us, Speaking to the World: Elizabeth Kolbert on the Craft of Environmental Journalism." *Edge Effects*, November 18, 2014. http://edgeeffects.net/elizabeth-kolbert/.

Dominus, Susan. "The Mystery of the Red Bees of Red Hook." *New York Times*, November 29, 2010.

Douglas-Hamilton, Iain, and Fritz Vollrath. "African Bees to Control African Elephants." *Naturwissenschaften* 89 (November 2002): 508–11.

Earth Observatory. "More Lawns Than Irrigated Corn." Accessed December 8, 2016. http://earthobservatory.nasa.gov/Features/Lawn/lawn2.php.

Ellis, Hattie. *Sweetness and Light: The Mysterious History of the Honeybee*. New York: Harmony Books, 2004.

Epstein, David, James L. Frazier, Mary Purcell-Miramontes, Kevin Hackett, Robyn Rose, Terrell Erickson, Thomas Moriarty, et al. "Report on the National Stakeholders Conference on Honey Bee Health." Report on the proceedings of the National Honey Bee Health Stakeholder Conference Steering Committee, October 15–17, 2012, Alexandria, Va. http://www.usda.gov/documents/ReportHoneyBeeHealth.pdf.

Fisher, Rose-Lynn. *Bee*. New York: Princeton Architectural Press, 2010.

Foucault, Michel. *Discipline and Punish: The Birth of the Prison*. London: Penguin, 1975.

Friends of the Earth. "Gardeners Beware 2014: Bee-Toxic Chemicals Found in 'Bee-Friendly' Plants Sold at Garden Centers Across U.S. and Canada." http://libcloud.s3.amazonaws.com/93/3a/3/4738/GardenersBeware Report_2014.pdf.

Frost, Robert. *Robert Frost's Poems*. Introduction by Louis Untermeyer. New York: Holt, Rinehart & Winston, 1971.

Garland, Max. "For a Dedication by the River." Poem written for the inauguration of Chancellor James C. Schmidt at the University of Wisconsin–Eau Claire, November 8, 2013. http://www.uwec.edu/inauguration/upload/Inauguration Poem.pdf.

Gil, Mariana, and Rodrigo J. De Marco. "Decoding Information in the Honeybee Dance: Revisiting the Tactile Hypothesis." *Animal Behaviour* 80 (2010): 887–94.

Golden, David B. K. "Insect Sting Anaphylaxis." *Immunology and Allergy Clinics of North America* 27 (May 2007). http://www.ncbi.nlm.nih.gov/pmc/articles/PMC1961691/.

Gough, Andrew. "The Bee: Part 2—Beewildered." July 2008. http://andrewgough.co.uk/articles_bee2/.

Grout, Roy A., ed. *The Hive and the Honey Bee*. Hannibal, Mo.: Standard Printing, 1946.

Hamill, Sam, trans. "Zazen on Ching-t'ing Mountain," by Li Po. Accessed November 14, 2016. https://www.poetryfoundation.org/poems-and-poets/poems/detail/48711.

Haraway, Donna. "Teddy Bear Patriarchy: Taxidermy in the Garden of Eden, New York City, 1908–1936." *Social Text* 11 (Winter 1984–85): 20–64.

Hirshfield, Jane. "Salt Heart." In *Lives of the Heart,* 14. New York: HarperCollins, 1997.

Hitchcock, Robert K. *Kalahari Communities: Bushmen and the Politics of the Environment in Southern Africa.* Copenhagen: International Work-Group for Indigenous Affairs, 1996.

Holmes, Richard. *The Age of Wonder: How the Romantic Generation Discovered the Beauty and Terror of Science.* London: Harper Press, 2008.

Horn, Tammy. *Bees in America: How the Honeybee Shaped a Nation.* Lexington: University Press of Kentucky, 2005.

Hubbell, Sue. *A Book of Bees.* New York: Ballantine, 1988.

Huseth, Anders S., and Russell L. Groves. "Environmental Fate of Soil Applied Neonicotinoid Insecticides in an Irrigated Potato Agroecosystem." *PLOS ONE* 9 (May 2014). http://journals.plos.org/plosone/article?id=10.1371/journal.pone.0097081.

Katcher, Joshua. "Art and Animals: An Interview with Giovanni Aloi." *Discerning Brute,* March 4, 2013. http://thediscerningbrute.com/art-animals-an-interview-with-giovanni-aloi/.

Kidd, Sue Monk. *The Secret Life of Bees.* New York: Viking, 2002.

Kleinman, Daniel Lee, and Sainath Suryanarayanan. "Dying Bees and the Social Production of Ignorance." *Science, Technology, and Human Values* 38, no. 4 (2013): 492–517. doi:10.1177/0162243912442575.

———. "Honey Bees Under Threat: A Political Pollinator Crisis." *Guardian,* May 8, 2013. https://www.theguardian.com/science/political-science/2013/may/08/honey-bees-threat-political-pollinator-crisis.

Kolbert, Elizabeth. *The Sixth Extinction: An Unnatural History.* New York: Henry Holt, 2014.

Kosek, Jake. "Ecologies of Empire: On the New Uses of the Honeybee." *Cultural Anthropology* 25, no. 4 (2010): 650–78.

Krulwich, Robert. "Cornstalks Everywhere but Nothing Else, Not Even a Bee." NPR, November 30, 2012. http://www.npr.org/sections/krulwich/2012/11/29/166156242/cornstalks-everywhere-but-nothing-else-not-even-a-bee.

LaJeunesse, Sara. "Diet Affects Pesticide Resistance in Honey Bees." Penn State News, November 3, 2014. http://news.psu.edu/story/332959/2014/11/03/research/diet-affects-pesticide-resistance-honey-bees.

Leopold, Aldo. *A Sand County Almanac: And Sketches Here and There.* Oxford: Oxford University Press, 1949.

Leskosky, Richard J. "Size Matters: Big Bugs on the Screen." In *Insect Poetics,* edited by Eric C. Brown, 319–41. Minneapolis: University of Minnesota Press, 2006.

Levenstein, Harvey. *Paradox of Plenty: A Social History of Eating in Modern America.* Oxford: Oxford University Press, 1993.

Lewis-Williams, J. D. "The Thin Red Line: Southern San Notions and Rock

Paintings of Supernatural Potency."
South African Archaeological Bulletin 36,
no. 133 (1981): 5–13.

Maeterlinck, Maurice. *The Life of the Bee.*
New York: Dodd, Mead, 1901.

Milesi, Cristina, Christopher D. Elvidge,
John B. Dietz, Benjamin T. Tuttle,
Ramakrishna R. Nemani, and Steven W.
Running. "A Strategy for Mapping and
Modeling the Ecological Effects of U.S.
Lawns." http://www.isprs.org/proceed
ings/XXXVI/8-W27/milesi.pdf.

Milesi, Cristina, Steven W. Running,
Christopher D. Elvidge, John B. Dietz,
Benjamin T. Tuttle, and Ramakrishna R.
Nemani. "Mapping and Modeling the
Biogeochemical Cycling of Turf Grasses
in the United States." *Environmental
Management* 36, no. 3 (2005): 426–38.

Morton, Timothy. "Guest Column: Queer
Ecology." *PMLA* 125, no. 2 (2010): 273–
82. https://pratthms101.files.wordpress
.com/2013/05/morton-timothy_queer
-ecologies.pdf.

Murong, Xi. "A Blossom Tree." http://www
.womenofchina.cn/womenofchina
/html1/people/writers/9/9156-1.htm.

National Gardening Association. "2013
National Gardening Survey." http://
garden.org/articles/articles.php?q
=show&id=3737.

National Pesticide Information Center.
"Pesticide Fact Sheets." December
2015. http://npic.orst.edu/npicfact.htm.

Nelson, Arnold L., and Eugene W. Surber.
"DDT Investigations by the Fish and
Wildlife Service in 1946." Chicago: U.S.
Department of the Interior, Fish and
Wildlife Service, 1947.

Nixon, Rob. "The Anthropocene: The
Promise and Pitfalls of an Epochal
Idea." *Edge Effects,* November 6, 2014.
http://edgeeffects.net/anthropocene
-promise-and-pitfalls/.

———. "Slow Violence and Environmental
Storytelling." *Nieman Storyboard,* June
13, 2011. http://niemanstoryboard.org
/stories/slow-violence-and-environ
mental-storytelling/.

———. *Slow Violence and the
Environmentalism of the Poor.*
Cambridge: Harvard University Press,
2011.

Oreskes, Naomi, and Erik M. Conway.
*Merchants of Doubt: How a Handful of
Scientists Obscured the Truth on Issues
from Tobacco Smoke to Global Warming.*
New York: Bloomsbury Press, 2010.

Pager, Harald. *Stone Age Myth and Magic
as Documented in the Rock Paintings
of Southern Africa.* Graz, Austria:
Akademische Druck-u. Verlangsanstalt,
1975.

Panchromos Limited. "Inscentinel Vasor."
Accessed October 8, 2016. http://www
.panchromos.com/stories/inscentinel/.

Parker, Bertha Morris, and Robert Gregg.
Insect Friends and Enemies. Basic
Science Education Series. Evanston, Ill.:
Row, Peterson, 1941.

Partap, Uma, and Tang Ya. "The Human
Pollinators of Fruit Crops in Maoxian
County, Sichuan, China: A Case Study
of the Failure of Pollination Services
and Farmers' Adaptation Strategies."
Mountain Research and Development 32,
no. 2 (2012): 176–86.

Penn State Extension. "Pesticides and Pollinators." June 11, 2015. http://exten sion.psu.edu/plants/vegetable-fruit /news/2015/pesticides-and-pollinators.

Perry, Caroline. "Robotic Insects Make First Controlled Flight." May 2, 2013. https:// www.seas.harvard.edu/news/2013/05 /robotic-insects-make-first-controlled -flight.

Pollan, Michael. "Border Whores." *Times* (London), March 9, 2002. http:// michaelpollan.com/articles-archive /border-whores/.

———. *The Botany of Desire*. New York: Random House, 2001.

Polycarpou, Lakis. "The Problem of Lawns." *State of the Planet Blog*, Earth Institute, Columbia University, June 4, 2010. http://blogs.ei.columbia. edu/2010/06/04/the-problem-of -lawns/.

Price, Jennifer. "A Brief Natural History of the Plastic Pink Flamingo." In *Flight Maps: Adventures with Nature in Modern America*, 111–66. New York: Basic Books, 2000.

Raffles, Hugh. *Insectopedia*. New York: Random House, 2010.

Ransome, Hilda M. *The Sacred Bee in Ancient Times and Folklore*. London: George Allen and Unwin, 1937.

Robbins, Paul. *Lawn People: How Grasses, Weeds, and Chemicals Make Us Who We Are*. Philadelphia: Temple University Press, 2007.

Schiavenza, Matt. "Dead Swine-Gate: Anatomy of a Chinese Scandal." *Atlantic*, April 2, 2013. www.theatlantic .com/china/archive/2013/04 /dead-swine-gate-anatomy-of-a-chi nese-scandal/274549.

Schiffman, Richard. "Mystery of the Disappearing Bees: Solved!" Reuters, April 9, 2012. http://blogs.reuters.com /great-debate/2012/04/09/mystery-of -the-disappearing-bees-solved/.

Scientific American. "How to Pick a Lawn Mower That's Easy on Man—and Nature." June 17, 2008. http://www .scientificamerican.com/article /how-to-pick-a-lawnmower/.

Scudder, Geoffrey G. E. "The Importance of Insects." In *Insect Biodiversity: Science and Society*, ed. Robert G. Foottit and Peter H. Adler, 7–32. West Sussex, U.K.: Wiley-Blackwell, 2009.

Seeley, Thomas D. *Honeybee Democracy*. Princeton: Princeton University Press, 2010.

Sewell, Anne. "Gold Miners All A-Buzz After Honey Bees Give Go-Ahead for Mining." *Digital Journal,* August 24, 2013. http://www.digitaljournal.com /article/356998#ixzz4CEJo7FLD.

Sheck, Laurie. "One Day." In *10 at Night: Poems*, 77. New York: Knopf, 1990.

Sinha, Indra. *Animal's People*. New York: Simon and Schuster, 2007.

Solnit, Rebecca. "One Magical Politician Won't Stop Climate Change: It's Up to All of Us." *Guardian*, May 15, 2015. http://www.theguardian.com/com mentisfree/2015/may/15/one-magical -politician-wont-stop-climate-change -its-up-us.

Spivak, Marla, E. Mader, M. Vaughan, and N. H. Euliss Jr. "The Plight of the Bees."

Environmental Science and Technology 45, no. 1 (2011): 34–38.

Steiner, Rudolf. *Bees.* Barrington, Mass.: Anthroposophic Press, 1998.

Stewart, Susan. *On Longing: Narratives of the Miniature, the Gigantic, the Souvenir, the Collection.* Durham: Duke University Press, 1993.

Thomas, Madeleine. "Climate Depression Is for Real: Just Ask a Scientist." *Grist,* October 28, 2014. http://grist.org /climate-energy/climate-depression -is-for-real-just-ask-a-scientist/.

Tsing, Anna Lowenhaupt. *Friction: An Ethnography of Global Connection.* Princeton: Princeton University Press, 2005.

UN Food and Agriculture Organization. "Crops, Browse, and Pollinators in Africa: An Initial Stock-Taking." 2007. First published by the African Pollinators Initiative in 2003. http:// www.fao.org/3/a-a1504e.pdf.

UN News Centre. "Humans Must Change Behaviour to Save Bees, Vital for Food Production—UN Report." March 10, 2011. http://www.un.org/apps/news /story.asp?NewsID=37731&Kw1=bees &Kw2=&Kw3=#.V3QcxdcWW3A.

Untermeyer, Louis, ed. *Modern British Poetry: A Critical Anthology.* New York: Harcourt, Brace, 1919.

U.S. Department of Agriculture, National Agricultural Statistics Service. *Vegetables: 2015 Summary.* February 2016. http://usda.mannlib.cornell .edu/usda/current/VegeSumm /VegeSumm-02-04-2016.pdf.

U.S. Environmental Protection Agency. "Carbaryl IRED Facts." Revised October 22, 2004. https://archive.epa.gov /pesticides/reregistration/web/pdf /carbaryl_ired.pdf.

U.S. Fish and Wildlife Service, Division of Environmental Contaminates. "Homeowner's Guide to Protecting Frogs: Lawn and Garden Care." July 2000. https://www.fws.gov/dpps /visualmedia/printingandpublishing /publications/2003_HomeownersGuide toProtectingFrogs.pdf.

Van Winkle, Dan. "Artist Uses Real Bees to Turn Figurines into Horrifying Beehive People." *Mary Sue,* February 21, 2014. http://www.themarysue.com /horrifying-bee-people/.

von Frisch, Karl. *Bees: Their Vision, Chemical Senses, and Language.* Ithaca: Cornell University Press, 1950.

Wargo, John, Nancy Alderman, and Linda Wargo. "Risks from Lawn Care Pesticides." North Haven, Conn.: Environment and Human Health, 2003. http://www.ehhi.org/reports/lcpesti cides/lawnpest_full.pdf.

Weisman, Alan. *The World Without Us.* New York: Picador, 2007.

Whitney, Kristoffer. "Tangled Up in Knots: An Emotional Ecology of Field Science." *Emotion, Space, and Society* 6 (February 2013): 100–107.

Williams, Raymond. "Ideas of Nature." In *Problems in Materialism and Culture,* 67–85. London: Verso, 1980.

Woolf, Virginia. *The Death of the Moth and Other Essays.* New York: Harcourt, Brace, 1942.

Wordsworth, William. *The Poetical Works of William Wordsworth*. London: Macmillan, 1896.

Interviews by Author

Adams, Sharon and Larry. Community activists, Walnut Way Conservation Corp., South Milwaukee, Wisconsin, May 23, 2013.

Bos, Colleen. Owner, Bos Meadery, Madison, Wisconsin, November 16, 2014.

Eisbach, John and Christina. Organic farmers, Galena, Illinois, August 10, 2013.

Farbing, Chen. President, Yingjing Honey Association, Yingjing, China, March 23, 2013.

Hartmann, Dennis and Shelly. Owners of True Blue Farms, Grand Junction, Michigan, July 17, 2013.

Isaacs, Rufus. Entomologist, East Lansing, Michigan, July 17, 2013.

Kdav, Arthur. Artist, author's father, Madison, Wisconsin, November 12, 2014.

Langston, Nancy. Environmental historian, Madison, Wisconsin, August 30, 2011.

Norman, Tyler. Member of Beehive Design Collective, Greater Madison, Wisconsin, August 15, 2010.

Stack-Whitney, Kaitlin. Staff biologist for U.S. EPA Office of Pesticide Programs, October 9, 2013.

Suryanarayanan, Sainath. Entomologist, Madison, Wisconsin, November 2, 2012.

Thiele, Michael. Owner of Gaia Bees, north of San Francisco, August 24, 2011.

Walker, Allen. Beekeeper, North Fort Myers, Florida, April 5, 2012.

Williams, Venice. Community activist, Alice's Garden, South Milwaukee, Wisconsin, May 23, 2013.

Woller, Eugene and Donna. Beekeepers and owners of Gentle Breeze Honey, April 4, 2011; October 3, 2012.

Xianlin, Yin. Beekeeper, Yingjing, China, March 23, 2013.

Ya, Tang. Professor of environmental biology, Sichuan University, Chengdu, China, March 25, 2013.

Credits

An earlier version of chapter 2 appeared as "Searching for the Bees of Guangxi and Sichuan," *Interdisciplinary Studies in Literature and Environment* 21, no. 4 (2014): 895–905. Permission for use granted by the Association for the Study of Literature and Environment and Oxford University Press.

An earlier version of chapter 4 appeared as "The Sorrow of Bees," *Aeon,* November 26, 2014.

An earlier version of chapter 7 appeared as "Guard Bee: Storying Resilience," *Resilience: A Journal of Environmental Humanities* 2, no. 2 (2015): 53–64. Permission for use granted by the University of Nebraska Press.

James Crews, "It Was Necessary" (unpublished), printed by permission of the poet.

Laurie Sheck, "One Day," from *10 at Night: Poems,* copyright © 1990 by Laurie Sheck. Used by permission of Alfred A. Knopf, an imprint of the Knopf Doubleday Publishing Group, a division of Penguin Random House LLC. All rights reserved.

Art by Sibylle Peretti reproduced by permission of the artist. Photography by Mike Smith (figs. 1.1–1.4).

Art by Rose-Lynn Fisher reproduced by permission of Fairbank Literary and the artist. http://www.rose-lynnfisher.com.

Art by Kim Gurney reproduced by permission of the artist. Sound recording and assistance by Brendon Bussy (fig. 3.2). Photography by John Hodgkiss (figs. 3.5–3.6).

Art by Sarah Hatton reproduced by permission of VAGA and the artist. Figure 4.1 © Sarah Hatton / CARCC, Ottawa / VAGA, New York. Figure 4.2 © Sarah Hatton / CARCC, Ottawa / VAGA, New York.

Art by Elizabeth Goluch reproduced by permission of the artist. Photography by Julian Beveridge (figs. 5.1–5.4).

Art by Aganetha Dyck reproduced by permission of the artist. Photography by Peter Dyck (figs. 6.2, 6.4).

Art by the Beehive Design Collective licensed under Creative Commons Attribution-ShareAlike 3.0. https://creativecommons.org/licenses/by-sa/3.0/.

Art by Lea Bradovich reproduced by permission of the artist.